The Rocky Mountains

MANAGING EDITORS
Amy Bauman
Barbara J. Behm

CONTENT EDITORS
Amanda Barrickman
James I. Clark
Patricia Lantier
Charles P. Milne, Jr.
Katherine C. Noonan
Christine Snyder
Gary Turbak
William M. Vogt
Denise A. Wenger
Harold L. Willis
John Wolf

ASSISTANT EDITORS
Ann Angel
Michelle Dambeck
Barbara Murray
Renee Prink
Andrea J. Schneider

INDEXER
James I. Clark

ART/PRODUCTION
Suzanne Beck, Art Director
Andrew Rupniewski, Production Manager
Eileen Rickey, Typesetter

Copyright © 1992 Steck-Vaughn Company

Copyright © 1989 Raintree Publishers Limited Partnership for the English language edition.

Original text, photographs and illustrations copyright © 1985 Edizioni Vinicio de Lorentiis/Debate-Itaca.

All rights reserved. No part of the material protected by this copyright may be reproduced or utilized in any form by any means, electronic or mechanical, including photocopying, recording, or by any information storage and retrieval system, without permission in writing from the copyright owner. Requests for permission to make copies of any part of the work should be mailed to: Copyright Permissions, Steck-Vaughn Company, P.O. Box 26015, Austin, TX 78755. Printed in the United States of America.

Library of Congress Number: 88-18337

2 3 4 5 6 7 8 9 0   97 96 95 94 93 92

Library of Congress Cataloging-in-Publication Data

Wingfield, John C., 1948-
   [Montagne rocciose. English]
   The Rocky Mountains / John C. Wingfield.

   — (World nature encyclopedia)
   Translation of: Montagne rocciose.
   Includes index.
   Summary: Describes the natural and ecological niches, boundaries, and life in the wildlife habitats of the Rocky Mountains.
   1. Mountain ecology—Rocky Mountains—Juvenile literature. 2. Biotic communities—Rocky Mountains—Juvenile literature. I. Title. II. Series. III. Series: Natura nel mondo. English. [DNLM: 1. Mountain ecology—Rocky Mountains. 2. Ecology—Rocky Mountains. 3. Biotic communities—Rocky Mountains.]
QH104.5.R6W5613   1988   574.5'.264'0978—dc19   88-18362
ISBN 0-8172-3325-3

Cover Photo: Lou Poulter—Colorado Tourism Board

WORLD NATURE ENCYCLOPEDIA

The Rocky Mountains

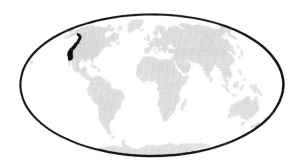

John C. Wingfield

Laramie Junior High
1355 N 22nd
Laramie, WY 82070

RAINTREE
STECK-VAUGHN
L I B R A R Y
Austin, Texas

CONTENTS

6 INTRODUCTION

9 FROM THE PACIFIC OCEAN TO THE GREAT DIVIDE

Mountains Rising from the Ocean, 9. The Cascades, 11. The Area Between the Mountains, 15. The Great Divide, 17.

21 THE OCEAN, THE COAST, AND PUGET SOUND

Seabirds, 21. Sea Mammals, 24. Stretches of Giant Seaweeds, 25. Sand Dunes and Brackish Marshes, 28. Puget Sound, 31. The Salmon, 33.

37 THE RAIN FORESTS

A Unique Phenomenon, 37. The Advantages of Conifers, 37. Various Species and Their Natural History, 40. Deciduous Trees and the Underbrush, 42. The Fauna: Small Animals, 44. Birds, 46. Mammals, 47.

53 MOUNTAIN FORESTS, ALPINE MEADOWS, AND PEAKS

Trees and Flowers, 53. Mammals, 55. Predators, 65. Birds, 72. Frogs, 75.

77 PLATEAUS

Huge Herds of Hoofed Mammals, 77. Prairie Dogs, Squirrels, and Hares, 79. The Skunks, 81. Predators, 82. Swans and Pelicans, 85. Common Cranes and Endangered Cranes, 89. Other Birds in Wet Environments, 90. Birds of Prey, 95. Grouse of the Sagebrush, 97. Reptiles and Amphibians, 99.

103 GUIDE TO AREAS OF NATURAL INTEREST

Canada, 105. United States, 108.

123 GLOSSARY

126 INDEX

INTRODUCTION

During the period from the sixteenth century to the eighteenth century, the first adventurous sailors began to land on the western coasts of North America. The first Europeans to reach the region were probably Russian hunters, looking for sea lion and sea otter furs. Other explorers, like captains James Cook and George Vancouver, landed on the northwestern coast and the Puget Sound area. They found a challenging and magnificent land, with huge mountains rising straight up from the sea. The mountaintops were white with snow and ice, and the slopes were covered with forests of giant trees. The first explorers to land on the western coast also found prosperous Indian populations, whose lives were mainly centered around salmon fishing. In the villages, they found "long houses" made of wood and impressive totem poles, carved with animals and gods of the ocean and the woods.

In 1792, Captain Robert Gray, an explorer sent by the United States government (the United States had just achieved independence), sailed up to the mouth of a huge river. The river was at the present-day border of the states of Washington and Oregon. The captain gave the river the name of his ship, *Columbia*, but he did not dare to venture upstream, toward the interior.

A few years later, two legendary explorers, Meriwether Lewis and William Clark, crossed the Rocky Mountains from the east and went down into this vast, wild area of mountain ranges, deserts, and grasslands. On their dangerous journey, the explorers followed the course of another large river, the Snake River, all the way to the Columbia River and then to the Pacific Ocean. Lewis and Clark were impressed by all they saw.

The first colonists arrived in the mid-nineteenth century. They were soon followed by a flood of settlers. Most of the pioneers wanted to reach California and Oregon, but many ended up settling along the northwestern coast and in the inland plains. The giant trees were cut down, and the grasslands were plowed or turned into grazing land for thousands of cattle and sheep. Today, although many of these wonderful natural environments have been tamed, some still remain for people to enjoy. Also, while some animals and plants have become rare or endangered, other interesting species are still common and widespread.

FROM THE PACIFIC OCEAN TO THE GREAT DIVIDE

The first European explorers to reach the extreme northwestern region of North America found a wild coastline, with high steep cliffs above the ocean. This type of coastline stretches from Vancouver Island in British Columbia (in southwestern Canada) all the way to California, a total of about 1,243 miles (2,000 kilometers).

Mountains Rising from the Ocean

The western border of the continent is entirely mountainous. South of Vancouver Island lie the isolated Olympic Mountains in the state of Washington, and then a long line of ridges and peaks of the Coast Range, ending with the Siskiyou Mountains in northern California. These mountains are seldom over 6,562 feet (2,000 meters) high and, unlike the inland ranges, are not of volcanic origin. Certainly though, they are a real geological puzzle—especially the Olympic Mountains.

Around seventy million years ago, the coastline started a process of uplift. The rocks being lifted were mainly sedimentary rocks which had been forming on the bottom of the ocean. The layers were pushed upwards, eroded, and then uplifted again.

Great heat and tremendous pressure caused changes in the structure of the sedimentary rocks. They became metamorphic rocks, mainly slates and phyllites. Within these rocks volcanic intrusions can still be found. These are volcanic rocks that are found between other rocks. These probably came from cracks opened on the bottom of the ocean, and were lifted up later on. The result is an incredible mix of sedimentary and metamorphic rocks, greatly folded and changed, and often even broken and shattered on the surface. This is due to erosion by frost, water, and wind.

More recently, during the ice ages, this entire area (except the higher peaks) was covered by an icy layer. Glaciers made their way down from the mountains, forming deep valleys. Around the end of the Pleistocene epoch of geological time (about ten thousand years ago), huge glaciers created wide valleys just east of the Pacific coastline. When the glaciers retreated, one of these valleys was flooded, creating Puget Sound. This deep channel is connected to the ocean by the Juan de Fuca Strait, which separates Vancouver Island from the Olympic Mountains. To the north, the delta of the Fraser River and the surrounding plains in British Columbia were formed. To the south,

Preceding page: A view of Glacier National Park in Montana shows an extremely beautiful park. This park is visited by about two thousand tourists every year. In the past it was a sacred area to the Blackfoot Indians. The park includes about fifty small glaciers, six large lakes, two hundred small lakes and glacial pools, a thousand waterfalls, and just about a thousand trails.

Opposite page: A stretch of rocky coastline is seen near Monterey, California. The western border of the American continent is lined with mountains.

The high cliffs of the rocky northwestern American coastline are only rarely interrupted by stretches of beach, but when this happens, the sandy ocean beaches are wide and spectacular. They often have tall dunes and many trees which extend to within a few feet of the ocean. In the picture, besides an arched sandy beach, there is also a rocky point in the background, with two large cliffs in front of it. They are the result of the eroding action of violent ocean waves, endlessly at work on the rocks.

Williamette Valley, in Oregon, was not flooded, because it was located at a higher altitude. Today, all these plains are very well suited for farming and are the most densely populated areas of the region. The cities of Vancouver, Seattle, and Portland were built on these plains.

The shores of the northwestern coast are constantly exposed to the powerful action of the ocean. To the west, water stretches almost 6,214 miles (10,000 km), while to the south it extends uninterrupted all the way to the Antarctic. Huge waves are caused by storms on the ocean. They strike

the coast with great power. Little by little, the rocky coast is eroded by the ocean. Water seeps into cracks and crevices. Wave action compresses the air and causes the rocks to literally explode, creating larger and larger openings. Also, the debris which has detached from the coast is thrown back and forth by the waves, helping the erosion.

During this erosion, successive "slices" of coastline break loose, with all their soil and vegetation. Huge tree trunks, up to 13 feet (4 m) long, are also thrown against the shore by the waves. Along most of the coast, the high tide level is marked by the rubble of boulders and tree trunks scattered here and there among the tide pools on the sandy or rocky beaches. The harder and more resistant rocks form isolated cliffs, which eventually break down into heaps and columns, separated from the main coastline. The high cliffs of the rocky northwestern coast only rarely have stretches of sandy beach. A good example of this is the wide, brackish (with a mixture of fresh and salt water) marshes of Gray's Harbor, in the state of Washington, and the Coastal Dunes National Monument on the southern coast of Oregon. Elsewhere, it seems that the mountains, often covered by mist and fog, rise straight up from the ocean.

The Cascades

To the east of the coastline and the large glacial valleys, an even greater mountain range rises. It is the Cascade Range. The Cascades emerge from the ocean in British Columbia and stretch south for about 1,243 miles (2,000 km) through the states of Washington and Oregon to Lassen Peak, in northern California. More to the south, the Cascades merge with the high Sierra Nevada of California and Mexico.

Like the mountain ranges along the coast, the Cascades were also lifted, folded, changed, and eroded by weathering and the action of glaciers during the last seventy million years. They are mainly made up of sedimentary rocks, but large volcanic intrusions are also found. Most of the peaks are no higher than 8,202 feet (2,500 m). Unlike the coastal mountain ranges, some of the mountains of the Cascades are volcanoes, rising in a line on top of the central "backbone" of the range, another 6,562 feet (2,000 m) high. These peaks are much younger (by about one million years) than the rest of the range, and most of them are still active volcanoes. Among them are Mount Rainier in Washington, and Mount Shasta in California.

The drawing shows what happened in the explosion of Mount St. Helens (May 18, 1980). After the explosion, the mountain was 1,312 feet (400 m) shorter. Later on, the crater filled many times with material which was then violently thrown out in an alternating cycle of construction and destruction.

Symmetrical, or evenly shaped, cones of these mountains were formed by volcanic eruptions, which spread many layers of lava onto wide areas. In recent times, the lava outflows have slowed down, and a thick layer of magma has collected inside the craters, forming a sort of "cork" of solidified lava. Very high pressures can exist beneath these lava corks, which eventually blow out in frightening explosions, throwing powdered rocks and hot gases high into the air. One of these explosions formed Crater Lake in Oregon. The volcanic ash covered most of the northwestern United States. In some places, the ash was several feet deep.

More recently, on May 18, 1980, the peak of Mount St. Helens in Oregon exploded, throwing masses of rock and ice into the air. By geological standards this is a relatively minor explosion, but its power has been estimated to be thousands of times greater than that of the atomic bomb exploded on Hiroshima. The lava from the volcanic crater flowed out over the mountain slope, killing over sixty people and millions of deer, bears, squirrels, birds, and so on. The flow also totally destroyed hundreds of square miles of primary forest (forest that had never been changed by humans). Ashes rose up to 9 miles (14.5 km) above the mountain and

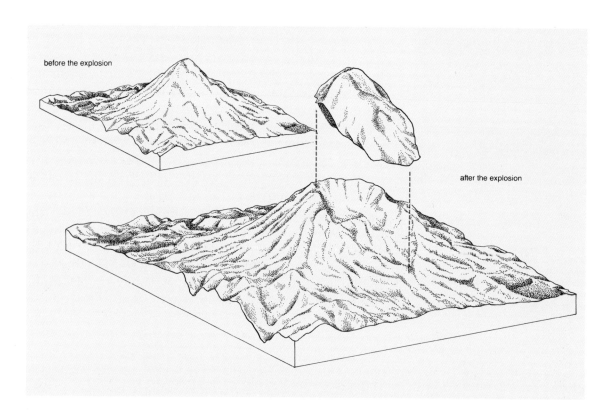

This picture shows the well-known Crater Lake, in the national park which is named after it, in Oregon. The Klamath Indians say that the lake and crater originated from the destruction of Mount Mazama, where the god Llao lived. The destruction was caused by the god Skell, who lived on Mount Shasta. The two gods are said to have fought over an Indian girl, and Skell won. According to geologists, a huge eruption, occurring around 6,800 years ago, weakened the structure of the volcanic cone of Mount Mazama. As a result, the mountain collapsed inward, taking on its present shape, with a rim 9,186 feet (2,800 m) high and a lake 2,296 feet (700 m) deep. A cone rising from the lake stands 820 feet (250 m) out of the water.

were carried all around the earth by powerful air currents, called "jet streams." In the central part of Washington state, the ashes fell so thick that rivers and lakes were clogged and some farmed areas were covered by a layer more than three feet deep.

Around the mountain, all life was completely destroyed, and the landscape was greatly changed. The once symmetrical peak, resembling a giant snow cone, had 1,312 feet (400 m) of its height blown off and showed an awesome crack on its western slope.

The heat melted snowfields and glaciers, and catastrophic mudslides flowed downhill, invading the plain up to 62 miles (100 km) west of the destroyed cone. In the middle of this devastating scene, a new lava dome has begun to grow on the old crater, and Mount St. Helens will probably regain its previous shape in the years to come as rocks from new eruptions are layered on top of each other.

A few years after this immense disaster, life had already begun to reappear in the areas around the volcano. Some rocks have been colonized by lichens, and the wind has carried many seeds onto the ash layers. Even small insects

These diagrams illustrate rainfall (above) and vegetation types at different altitudes (below) for a stretch of land from the Pacific Ocean to the Great Plains. *Above:* The solid line represents the yearly rainfall; the dotted line the rainfall during the dry period (April through September). Note that rainfall decreases to the east of the Sierra Nevada and Rocky Mountains.
Below: R = redwood; SP = pine; BG = bunch grass; C = chaparral; S = spruce; J = juniper; SB = sagebrush; P = Douglas fir; A = alpine grassland; SG = short grass.

and spiders, perhaps also carried by the wind, hide in the safest spots among the pioneering vegetation. The first seedlings of the giant conifers (trees with cones) are now sprouting, and deer have come back, venturing around the foothills of the mountain. Perhaps a large primary forest will grow back in a few hundred years, and perhaps it will someday be destroyed again by the mountain. This cycle of destruction and regrowth has probably occurred many times around the volcanic peaks of the Cascades.

Unlike the violent method of the volcanoes, ice has smoothed and polished peaks and ridges for many millions of years. Several glaciers still exist today, built by deep snowfalls. The never-ending action of ice and fire has created a maze of shattered rocks, snowfields, glaciers, alpine meadows, and forests. The Cascades are broken only in one place, where they are crossed by the powerful Columbia River, marking the Washington-Oregon border. The many torrents and rapids which the river formed as it cut through the mountains are all gone. These were flooded by an artificial lake which was formed by a dam. But the name *Cascades* is still appropriate for the range. A great many streams and torrents come tumbling downhill, formed by the thawed waters of snowfields and glaciers.

West of the Cascades, the climate is mild and humid. Along the coast and on the plains, temperatures rarely rise

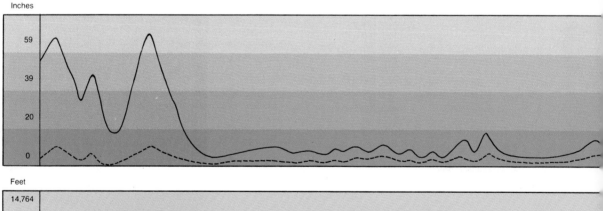
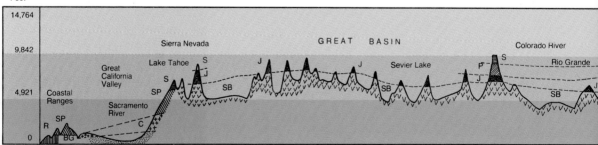

above 86° Fahrenheit (30° Celsius) during the summer, and rarely drop below the freezing point in winter. The amount of rainfall can be very high. The humid air coming from the ocean rises along the slopes of the coastal ranges and the Cascade Range, cooling down and condensing in the process. As a result, rainfall along the coast can be as high as 315 inches (800 centimeters) a year.

At higher altitudes, winters are colder and most of the precipitation comes in the form of snow. Some years, the snow forms a layer as deep as 49 feet (15 m) or more. One winter, over 131 feet (40 m) of snow fell on Mount Rainier. As a result, wide snowfields are formed. They do not melt completely during the summer and help to form glaciers.

The Area Between the Mountains

Moving east, away from the Cascade Range, the climate gets drier and drier, because the clouds have already dropped most of the rain on the western mountain ranges. The eastern basin and the plains which follow it are located in the "rain shadow" of the western ranges (the mountains shadow them from the rain). Heading down toward the Columbia River, pine trees, junipers, and oaks gradually take the place of spruces and firs. Wide grasslands are also found. In the lower areas, close to the river and on the plateau of central Oregon, the grass cover gets thinner and broken. Here the main type of vegetation is sagebrush. In some areas, sand dunes and lava flows can also be found. Rainfall in this region is never higher than 20 inches (50 cm) a year, and in some areas it is much lower than that. Summer temperatures are often higher than 104°F (40°C), while winter temperatures can drop to -4°F (-20°C). One winter during the 1930s, temperatures on the slopes of the Rocky Mountains fell to -92°F (-69°C).

Geologically, the area between the mountains is very complex. It includes the basin of the Columbia River (from southern British Columbia to central Washington), the canyon of the Snake River, the Palouse region (Idaho and central Washington), the plateau of central Oregon, and the Great Basin Desert (Utah, Nevada, and southern Oregon). The region is also strewn with forty-five isolated mountains, which form a sort of chain stretching from British Columbia all the way down to Mexico.

One hundred million years ago, a geological upheaval created the metamorphic rocks and the granite mountains which rise, here and there, from the plain. Much more

The drawing shows the typical look of columnar basalt. It originated directly from the rapid hardening of a lava flow. Spectacular basalt of this kind, usually with hexagonal columns, can be found not only in the Rocky Mountains region, but in most volcanic areas on earth.

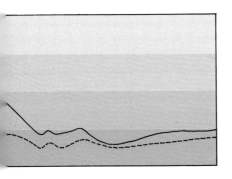

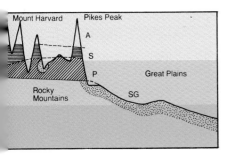

The Grand Teton Range, in Wyoming, is a mountain range rising up to over 13,123 feet (4,000 m). It is the heart of the national park which is named after it. In some places the original rocks are covered by sedimentary layers. These in turn, are covered by lava. Elsewhere, various rock formations rise skyward, free of all sedimentary rock.

recently, about three million years ago, molten lava flowed out of cracks up to ½ mile (1 km) long, covering most of the basins of the Columbia and Snake rivers. In some places, the lava layer is 4,921 feet (1,500 m) high. In other places, unusual lava formations are found. As the lava cooled off, it fractured into symmetrical columns up to 98 feet (30 m) tall.

The many rivers which flowed out of the glaciers surrounding the region eroded the lava flows and created ravines and deep gullies, or coulees. Even today, many torrents flowing down from the mountains seem to disappear inside the porous volcanic rock, only to come out again as springs somewhere else. The Columbia and Snake rivers have cut basins and canyons throughout the region. The Snake River, especially, flows through a wild region of mountains, deserts and sagebrush, from the Rocky Mountains to the Columbia River.

The Great Divide

In the easternmost areas of the region, the mountain ranges become more and more frequent. They eventually join to form the largest chain in North America: the Rocky Mountains. The Rockies divide the continent in two, with the drainage basins of the Columbia, Snake, and Colorado rivers to the west. These rivers flow into the Pacific Ocean.

On the other side of this range, the Missouri, Platte, and Arkansas rivers flow down the eastern slopes of the mountains and make their way to the Gulf of Mexico. The mountains which divide the eastern and western drainage basins are called the "Continental Divide." This is a very wide region, with about sixty mountain ranges. It stretches from Canada to Mexico, and the land is extremely varied. High rock walls and ridges alternate with forests, deserts, grasslands, and grassy areas in the mountains known as "parks."

Incredible "cascading" terraces of travertine are found in Yellowstone National Park, Wyoming. Travertine is a mineral deposit laid down from the hot spring waters. When the water cools off, the alkaline salts in it are deposited, and little by little build these magnificent natural "sculptures."

The Rocky Mountains were formed about five million years ago. Like other western mountains, they are mainly composed of sedimentary and metamorphic rock layers, along with granite intrusions. On these mountains, rain and snowfall are abundant. This is because moisture cools and condenses as air rises up the slopes. During the winter, up to 197 feet (60 m) of snow have been recorded on the peaks, and there are many glaciers.

During the glaciations of the Quaternary period ice tongues invaded the plains. Today, they have withdrawn to the high peaks, where snow accumulates on the side away from the wind. A depression shaped like a large bowl, called a "cirque," is created. Here glaciers are also formed. By erosion they widen the cirques. Sometimes, cirques merge together, forming sharp and steep dividing ridges where they join. These ridges are called "knife blades" or "arretes." As the glacier continues to erode the rock, the knife blades eventually collapse and a mountain pass is formed. When the glacier melts, a lake can form on the bottom of the cirque. Water is kept in the hollow mainly by narrow ridges of rubble which have been left behind by the glacier itself. These ridges are called "moraines." The high Rocky Mountains, like other large mountain ranges, are a mixture of cirques, glaciers, lakes, snowfields, and alpine meadows.

Springs of hot and mineral waters are common in the Rocky Mountains. The water is heated by pressure from magma lying deep below the surface. Some of the most spectacular springs are in Yellowstone National Park, in Wyoming. Here, water reaches temperatures well over the boiling point, and at varying times, hot steam blows out from openings in the ground, called "geysers." Some geysers, like the one called "Old Faithful," erupt very regularly, about every sixty minutes. Others are more irregular and may erupt many times in a few minutes, then remain silent for hours. Most of these geysers are quite spectacular, and send steam clouds many feet into the sky.

The entire region has much variety and many contradictions. To the west grow moist forests, soaked with warm rain from the Pacific Ocean. Inland, there are dry lava flows, with almost no life. Then there are wide areas of grassland and sagebrush stretching between the towering snow-covered peaks. It is not surprising that the Rocky Mountains are of such interest to biologists.

THE OCEAN, THE COAST, AND PUGET SOUND

The journey across the northwestern Pacific regions begins with the ocean. Due to a current coming from the north, the water along the coast is generally cold. In summer, though, the warm Japanese current often reaches the shores. This current, after crossing the ocean from west to east, heads north up to the coasts of Washington and British Columbia.

Along most of the coast, the continental shelf is relatively narrow, extending out for about 50 miles (80 km). Water rises up from the bottom of the ocean, carrying a lot of nutrients to the coastal waters. For this reason, plankton is plentiful. Plankton is a mixture of tiny or microscopic marine organisms. Larger crustaceans and their larvae feed on plankton, and are in turn eaten by fish or by other invertebrates, like squids. The smaller fish are eaten by larger fish, like salmon, rock fish, blackfish, and sea bass.

Seabirds

Many seabirds take advantage of the abundance of fish. They often nest on the rocky outcrops and small islands along the coast. Cormorants are particularly common. They feed mainly on small fish and build their nests on the most difficult to reach rocky coasts. The nests are large and made with twigs and seaweeds which the birds have gathered along the shore or in the water. They are always built on a small projection or on a cliff, often very close together. In April and May, the cormorants lay up to six chalky-white eggs. The newly hatched birds are completely featherless and helpless. They are totally dependent on their parents' care and will spend at least two months in their nest before they are able to fly down to the ocean.

Two other species commonly found nesting in this region are the common murre and the tufted puffin. They both belong to the Alcidae family. When floating on water, they resemble little penguins, but they are not related to these birds. The Alcidae family of birds is easily distinguished from penguins because they are all able to fly, while penguins cannot. Also, penguins live only in the Southern Hemisphere, while the Alcidae live in the Northern Hemisphere. Common murres and tufted puffins in the northern oceans occupy the same ecological niche that the penguins occupy in the southern oceans.

Both common murres and tufted puffins can dive for small fish and invertebrates. Common murres nest in large, crowded colonies on the tops of isolated cliffs along the

Opposite page: Two California sea lions and a gull bask on the craggy coast of the Pacific Ocean. Seals and sea lions are common along the western coasts. At a site north of Florence, Oregon, the sea lions gather in large groups and reproduce in huge sea grottos. They stay in the area for over six months.

Tufted puffin

coast, or on high, steep, and rocky outcrops. During their mating season, their grunts and whinings can be even louder than the noise of the ocean waves breaking on the shore. Huge amounts of their droppings collect on the cliffs, making the colonies easy to spot from a distance. The dried droppings, called "guano," are white due to the uric acid they contain. They cover the entire area.

Common murres lay a single egg on bare rock, with no attempt at building a nest. The eggs are rather pointed at one end. Any accidental push, which often happens in such a densely populated colony, will make the egg roll around in a circle. Its shape decreases the chances of it falling from the cliffs.

Tufted puffins, on the other hand, nest in holes they dig in the ground, usually on top of a little island, or sometimes on a high coast. At the beginning of the mating season, the puffins can be seen perched at the entrances to their nests. After the egg has been laid, both male and female take turns sitting on it. While one broods the egg, the other goes out to sea to catch food.

Fork-tailed petrels and Leach's petrels also make their nests in holes that they dig in the ground. These birds are very tiny, barely larger than a sparrow. They spend most of

Opposite page, above: A group of murres wear their spring plumage. When the mild weather arrives, these birds gather in large numbers on isolated cliffs and high rocks to lay their eggs. In early summer, when the young are ready to fly, the murres head back to the open sea, where they will remain until the following spring.

Below: A group of herring gulls and Heermann's gulls rest on a sandy stretch of beach on the western coast of the United States. Among the herring gulls, the adults are white with gray wings, while the young are marbled gray. Among the Heermann's gulls, the adults are gray with a black-and-white tail, while the young are an even brown.

their lives out on the open sea. They come ashore only to nest. Often, petrels fly for hundreds of miles to get their food, so the nestling may be left alone in the nest for several days. Both eggs and nestlings, though, have developed a resistance to cold. The nestlings can even lower their body temperature and enter into a kind of numbness, similar to a mammal's dormancy, or hibernation, as they wait for the adults to come back with food. In this way, they can save energy, which would be used to keep their body temperature at its normal level.

Perhaps the most common birds along the Pacific coast are the western gull and the glaucous gull. Both species usually nest on steep, rocky coasts, on points of land, and on little islands. They look for food along the shores, in tide pools, or on the ocean surface. They eat garbage and steal eggs and nestlings from other birds' nests. In some bird colonies, like those of murres, terns, and petrels, such predation can have disastrous effects and cause the loss of many eggs and nestlings, and sometimes even adults.

Both species of gulls lay three eggs in a nest which they make with grass, roots, seaweeds, and even with paper or pieces of rope. Usually, the nests are grouped in colonies,

sooty shearwater

black-footed albatross

northern fur seal

California sea lion

but they may also be isolated. The newly hatched birds are covered by a gray down, with black spots. This camouflages them and also serves as good protection against the cold ocean winds and the rain.

Not all of the seabirds that are found along the Pacific coast nest there. Many shearwaters, for example, sooty-shearwaters, spend the summer on the ocean of the continental shelf, feeding on fish, squids, and other invertebrate animals. These birds nest during the North American winter on the islands surrounding Australia and New Zealand, where it is summer in the Southern Hemisphere. They molt (shed their feathers) while they are in the north. In May, when they have just arrived, and in September, when they are getting ready to leave, it is easy to see them gathered in huge flocks, sometimes numbering in the thousands. Another rather common bird is the black-footed albatross, which is quite a large bird. Its wingspan is often over 6 feet (2 m). This albatross nests on the tropical islands of the Pacific Ocean, and at times other than its mating period, it comes close to northwestern America to feed on material thrown overboard by fishing boats. It is actually quite easily lured close to a boat by throwing fat or fish bowels out to sea.

Sea Mammals

Many species of mammals live along the coast. Among the various whales which pass by this area during their migrations, the most commonly seen is the gray whale. It migrates from its reproducing area, near Baja California (in Mexico), to its summer dwelling area in the Bering Sea, west of Alaska. Most of the cetaceans which are members of a group of marine mammals including whales, dolphins, and killer whales, do not actually live near the coast of western North America. They are just seen in passing.

Sometimes northern fur seals and the California sea lion can be seen, but usually they stay well off the coast. Among the seals and sea lions, only the Steller's sea lion reproduces along stretches of coastline that are most exposed to waves. This is a truly huge animal. The males can be up to 13 feet (4 m) long, and weigh some 1,984 pounds (900 kilograms). The females are much smaller, not longer than 8 feet (2.5 m) and weighing up to 992 pounds (450 kg). These animals feed mostly on fish and mollusks, especially herring and squids. They can dive down to 590 feet (180 m) looking for their prey.

Due to its fish-based diet, the Steller's sea lion has been

Sea lions, gulls (white), and cormorants (dark), rest on rocks along the western coast. The Pacific coastline is the home of many species of seabirds and mammals which feed on the rich food resources of the ocean. Sea lions feed mainly on fish and mollusks and can dive very deeply, even 590 feet (180 m), looking for their prey.

brutally hunted by fishermen, who accuse it of robbing nets loaded with fish. Today, this species is most often found in the open sea. However, they are often found in Puget Sound and sometimes venture upstream in large rivers like the Columbia River. When on land, this animal is quite suspicious and distrustful. Large groups of them can be seen on the southern Oregon coast at the Sea Lion Caves.

During the mating period, dominant males protect entire harems of females and often engage in violent and noisy clashes against rivals who either do not have a harem or rule over nearby territories.

About one year after mating, the females give birth to one baby, nursing it on land for three months. Then they go back into the ocean with the offspring.

Stretches of Giant Seaweeds

In many areas along the coast, including Puget Sound, giant seaweeds called "kelp" form wide, underwater "forests." These provide excellent shelter for many invertebrate animals, fish, and seabirds. The kelp seaweed can be some 66 feet (20 m) long. During its fast-growing period, it is said to be able to grow 10 inches (25 cm) a day! Thus, it easily

The head of a sea otter emerges from a bed of seaweed. This is a large carnivore belonging to the Mustelidae family. It is adapted to living in the ocean as well as, or even better than, seals, even though its paws are not fin-shaped, like a seal's. Sea otters, instead of resting on the rocks as seals and sea lions do, sleep in the seaweed beds and wrap themselves in long strands of seaweed.

comes up to the sea surface from the bottom where it is rooted, forming a kind of a floating mat.

One of the most interesting inhabitants of these seaweed mats is the sea otter, a carnivore similar to the better-known and more common river otter. The sea otter was almost extinct, due to heavy hunting for its fur. Today it is strictly protected, and its populations are starting to grow again. It is found mainly off the coast of California, but small groups have been released along the coasts of Washington and Oregon.

The sea otter is beautifully adapted for life in the ocean. It can be 5 to 6 feet (1.5 to 2 m) long, and its hind paws are large and webbed. Its fur is dark brown, mixed with gray, especially on the head and shoulders. It is very thick, and its hairs are about 0.8 inches (2 cm) long. When the otter dives into the water, the thick fur holds a certain amount of air, providing insulation from the cold. Sea otters spend much time carefully and thoroughly cleaning themselves. Dirty or tar-stained fur would completely lose its insulating properties, and the animal would quickly die of cold.

Normally, sea otters never come to shore, but instead spend all their lives around mats of seaweed. Here they swim on their backs, clean themselves, and sleep, after wrapping themselves in the long seaweed, which they use as safety belts. They feed mostly on large mollusks and sea urchins, and sometimes on fish and crustaceans. They are also known to be among the few animal species which use tools. They put a rock on their chest, and use it to crack open the sea urchins.

Sea otters can give birth any time of the year, usually to one young, or more rarely to two. The young stay with their mother for about a year.

Sand Dunes and Brackish Marshes

Low, marshy beaches are not common along the Pacific coast. Rocky cliffs, however, are most common. Some good examples of such environments, though, are Gray's Harbor and Leadbetter Point in the state of Washington, and the Coastal Dunes National Monument in Oregon. Also, along Puget Sound there are wide, brackish marshes near the delta of the Fraser River in British Columbia, and along the lower Skagit River in Washington. In these areas, for many square miles, large stretches of sedge and glasswort plants cover the soil. The marsh is cut through by twisted channels and interrupted by wide expanses of silt.

Opposite page, from top to bottom: A sea otter collects a rock from the sea bottom, carries it to the surface and uses it, balanced on its chest, as an "anvil." On the rock, it batters and breaks open the shells of mollusks, which are part of its diet. When on the surface, this clever animal rests on the floating seaweed, wrapping itself in the long strands to avoid drifting away. It moves slowly, always swimming on its back, playing in the waves with other otters or carrying young close to its chest.

Below: A knot (in its winter plumage) looks for food along a California beach.

At high tide, these silt banks are at least partly flooded.

Above the tide zone, sand dunes up to 164 feet (50 m) high can form. They are often colonized by sitka spruce, either alone or together with rhododendrons, salal, Oregon vine, and brambles. Farther inland, there are freshwater pools with sphagnum moss, the carnivorous (meat-eating) round-leaved sundew, cranberry bushes, and Labrador tea. Around these marshes grow thick woods of sitka spruce and other conifers, at times mixed with small groups of maples and alders.

Brackish marshes and silt banks are an ideal refuge for the migratory birds which stop here looking for small crustaceans, worms, and mollusks. During their winter migration, huge flocks of knots gather by the thousands on the silt banks, looking for food. The greater and lesser yellowlegs have very long bills, and they can search deep in the mud. Other species, like the dowitcher and the long-billed curlew, have extremely long and curved bills, so they can reach down even deeper.

On the beach, small groups of sanderlings look for food, precisely on the line which divides the shore from the water. From far above, they look like frantic little spots. They sprint forward following a retreating wave, then dart back to avoid the next breaker crashing onto the shore.

On the rocky points are found still other species of shorebirds. The ruddy turnstone and the black turnstone

pigeon hawk

actually turn stones over, looking for worms and crustaceans. The surfbird, rock sandpiper, and wandering tattler peck at the rocks covered with barnacles. Past the sand dunes, in the freshwater pools, many bird species can be found including herons and ducks.

The shores and marshes are home to some very efficient winged predators. Two species of hawks, the pigeon hawk and the peregrine falcon, usually hunt or prey on the shorebirds. On a sunny autumn afternoon, one might witness scenes such as the following. While out hunting, a peregrine falcon frightens a flock of red-backed sandpipers off a mud bank. The flock rises, staying tightly grouped, turning and diving as if a single bird. The birds' movements are so exactly synchronized that with every turn the color of the flock changes, depending on whether the birds are showing their pale chests or their dark backs. By the way, no one has yet been able to understand how these birds can

Opposite page, below: The rough-legged hawk is so called because of its completely feathered legs and toes. It lives in cold areas all over the Arctic, and in Europe and Asia. Its color can vary. In the picture, the bird is in its so-called light-colored stage.

achieve such perfect harmony and extreme synchronization. The peregrine falcon, though, attacks with great determination, flying swiftly and directly below the frantic flock. With a darting sprint upward, the predator captures its prey. With a half somersault, it strikes one bird. The victim falls dead onto the water. The falcon dives, catching the bird in flight, and makes its way back to its perch near the marsh. There, it starts to eat its meal at ease.

In winter, many shorebirds leave the area for warmer regions in Central and South America. Some, though, stay and spend the winter, often in large numbers. Among these winter dwellers, there are five species of ducks, the blue goose, and the whistling swan. The birds of prey also spend the winter here, and in some winters various species of hawks, eagles, and nocturnal birds of prey are found in great numbers. Both the pigeon hawk and the peregrine falcon stay over winter, and in some years even the snow-white gyrfalcons come down from the north. Often, the marsh hawk can also be seen, flying low near the marshes and reeds in search of prey. It mainly feeds on small rodents.

The red-tailed hawk scans its surroundings perched on poles or trees, while the rough-legged hawk glides like a huge moth, looking for small rodents.

Near brackish waters, or along the coast, it is not rare to spot a bald eagle perched on a tree or a rock, waiting for dead fish or other animals to be washed ashore by the tide. Some years, the region is literally invaded by snowy owls, which can be seen perched on the flat ground or on high places in the terrain. They are waiting for the right moment to pounce on a duck, a coot, a rabbit, or any other animal that would make a good meal.

The quietest spots in Puget Sound are also the home of seals. During the hottest hours these animals often come ashore and lie on the rocks or sand, basking in the sun. Seals can be 6 feet (2 m) long, and are usually grayish with lighter spots. Unlike sea lions, which can use their front fins to move on the ground, seals are not able to do so. For this reason, when on land, they have to move by crawling with an undulating, or wavelike, motion, much like a caterpillar or giant larva. They feed on fish and marine invertebrates, and mate in September.

Puget Sound

The seemingly calm waters of Puget Sound are really deceiving. High tide brings strong currents, and violent

Similar to a huge two-colored toy, the head of a killer whale emerges on the surface of the Pacific Ocean. This is one of the most amazing marine mammals, and also one of the most ferocious. Killer whales hunt not only fish and birds, but also seals and whales. On the other hand, they do not show any aggression toward humans. They grow up to 30 feet (9 m) long and are easy to recognize by the large white spots behind their eyes (partly visible in the picture) and by the tall, pointed dorsal fins, which can sometimes stand over 3 feet (1 m) tall.

whirlpools originate in deep places. At some points, water is as deep as 984 feet (300 m). In spite of this, in winter huge numbers of diving ducks stop here. Often, even killer whales venture inside this deep fiord which is a narrow valley flooded by the sea. There, they hunt salmon and sometimes also seals and seabirds. The killer whale is a mammal, related to dolphins. It can be up to 30 feet (9 m) long, and usually hunts in small groups, called "pods." It is easily recognizable by its triangular dorsal fin, which sticks out of the water as it swims just below the surface. As might be expected of a predator hunting large marine vertebrates, the

killer whale can swim very fast, perhaps up to 25 miles (40 km) per hour.

In Puget Sound, as well as in other places along the coast, is a peculiar bird, very interesting and mysterious. It is the marbled murrelet, a small bird of the Alcidae family. Its plumage is dark brown in summer and black and white in winter. It is usually seen in pairs, diving in the water to catch small fish and marine invertebrates. Up to a very few years ago, the marbled murrelet was the only bird nesting in North America whose nest had never been found. In 1978, much to their surprise, scientists discovered that this bird does not nest on the coast, but in the middle of forests. It builds its nest among moss and lichens, in high branches of conifer trees. After the hatching of the egg, the young, which is soon able to take care of itself, leaves the nest and cautiously lands on the forest ground. Then it makes its way to the nearest stream and swims along it all the way to Puget Sound or to the open sea, carefully followed and fed by its parents.

The Salmon

One of the most peculiar natural happenings of the Pacific coast is the salmon migration. These fish migrate from their places of hatching, in fresh water, to the ocean, and then back to the streams, to lay their eggs.

The Indians of the western coast, for example the Tlingit and Haida Indians, believed salmon to be supernatural beings who spent a long time in the ocean, could turn into human beings, and could dance and party. In spring and autumn, they would turn into fish and swim up the rivers to lay their eggs. Then, some of them would sacrifice themselves to the Indian fishermen. After their deaths, the Indians believed the salmon spirits went to their "Large House" on the bottom of the sea, and waited to reincarnate the following year.

Unfortunately, the ancient Indian legends could never have foreseen the efficient fishing methods of modern man. Today, salmon fishing, once so important in this region, is sharply declining. Salmon are caught on their way back to the rivers, and even in the ocean depths. Their populations have greatly dropped in the last few years, and all attempts at releasing some artificially hatched salmon in the rivers have not improved the situation significantly.

Five different species of salmon live in the Pacific Ocean. All of them are anadromous, which means they

Marbled murrelets are shown in their spring plumage (left) and winter plumage (right). These seabirds are especially interesting because their reproductive habits are unusual and were discovered only a few years ago. Their adaptation to nesting in trees, and not on rocks, is made possible because of the wide expanses of conifer forests along the American western coasts.

Below: The five species of Pacific salmon are shown as they swim upstream. From left to right: male chum, female and male sockeye, coho, humpback, and chinook salmon (all males).

Opposite page, above: Lamprey are among the worst enemies of salmon. They are primitive vertebrates. Their mouths do not have jaws, but instead they are sucker-like. This allows the lamprey to cling to the salmon and feed on their blood and tissues. Lamprey, like salmon, migrate from the ocean into fresh waters to spawn. The young spend their larval stage buried in a short tunnel on the bottoms of streams and eventually make their way back to the ocean, where they will reach maturity.

migrate from seawater to fresh water to reproduce. The young go back to salt water, and they grow to be adults in the ocean. Some species start swimming up the rivers in autumn, while others, such as the chinook salmon, can start their journey at any time.

The migrations of some species can be very short, but some other species, like the sockeye salmon, can travel for over 1,864 miles (3,000 km) and rise in elevation up to 2,296 feet (700 m).

While swimming up the rivers, these fish can swim against currents as strong as 3 miles (5 km) per hour and perform incredible leaps over rapids and waterfalls. Lakes encountered along the way provide food and a chance to regain strength lost while battling rushing currents in the upstream journey. These lakes are also split-off points to different egg-laying areas, located along various streams.

During their migration, salmon reach sexual maturity and their color changes. The male red or sockeye salmon turns from steel gray to bright red and develops a large hump on its back and peculiar hooked jaws. The females also turn reddish, but they are not as bright as the males, and do not undergo major changes in shape or structure.

During their migration, some salmon use 90 to 96 percent of their fat reserves, and from 30 to 50 percent of their body flesh. As they swim in shallow waters or try to get over

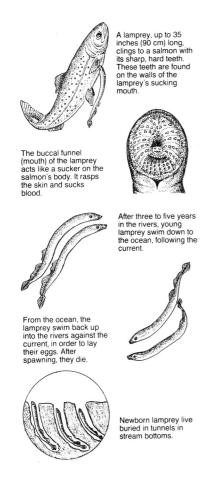

A lamprey, up to 35 inches (90 cm) long, clings to a salmon with its sharp, hard teeth. These teeth are found on the walls of the lamprey's sucking mouth.

The buccal funnel (mouth) of the lamprey acts like a sucker on the salmon's body. It rasps the skin and sucks blood.

After three to five years in the rivers, young lamprey swim down to the ocean, following the current.

From the ocean, the lamprey swim back up into the rivers against the current, in order to lay their eggs. After spawning, they die.

Newborn lamprey live buried in tunnels in stream bottoms.

rapids, they are easy prey for bears, bald eagles, and humans.

The fish, which eventually reach their reproduction areas, form pairs. The female makes a nest, called a "redd," on the bottom of the stream, digging a sort of hole in the gravel with sideways movements of its tail. Then, the complex courtship ritual takes place, ending with the laying and fertilization of the eggs in the redd.

Between 700 and 1,000 eggs can be laid in a single redd, and a female can lay 1,500 to 2,000 eggs, depending on its size and the species to which it belongs. The eggs are covered with gravel, allowing the oxygen-rich water to circulate around them, but preventing exposure to light which would be fatal to the developing embryos. After mating and laying eggs, all adults die, and streams and lakeshores are suddenly full of rotting carcasses. These attract large numbers of scavengers, like the common raven and the bald eagle.

The eggs hatch after fifty days to five months, depending on the temperature in the environment. After hatching, the young fish, called "fry," stay in the gravel for three to five weeks. When they come out, they are called "parrs." The parrs begin to swim downstream, toward the lakes and the ocean. Some will stay in a lake for about a year, feeding on crustaceans, while others may soon start their migration.

Shortly before the migration starts, salmon of all species begin gathering in groups and swimming with the current. In calm waters they will actively swim, while in the rapids they will let the current carry them, tail first. During the migration, some species undergo a real metamorphosis, or change of shape, reaching their ocean-going shape. At this stage, the salmon are called "smolts." Once in the ocean, the smolts can stay near the coast for a few months or up to a year. Then they move toward the open ocean, up to 994 miles (1600 km) off the coast, and stay there until they are near maturity. Salmon of the Pacific Ocean, depending on the species, may spend one to eight years in the ocean before returning to fresh waters, swimming up the rivers as mature adults, reproducing and dying.

Following the industrial development of the northwestern Pacific region, many barriers that the salmon cannot cross have been built along the rivers. These include dams on large rivers, like the Columbia River. Only a few dams are equipped with "fish ladders." Fish ladders are a series of higher and higher pools with which these fish can bypass the dam.

THE RAIN FORESTS

From California all the way up to Alaska, the coast, valleys, and the western slopes of the Cascades are covered by forests of giant conifers. About twenty different species of trees, belonging to eleven genera (groups of species) are found here. Some of them can be as much as 36 feet (11 m) in diameter and over 262 feet (80 m) tall. These are not only the largest organisms ever to have lived on earth, but they are probably also the oldest. Many of these trees are at least 1,000 years old, and some can even be as old as 3,500 years.

A Unique Phenomenon

The dominance of conifers found in the forests of these regions is a unique phenomenon, not found elsewhere at the same latitude in the whole Western Hemisphere. Usually, deciduous forests or mixed forests prevail. It is not known for sure how the conifers have come to dominate the environment. During the Miocene epoch (which took place fifteen to thirty million years ago), most of the Pacific coast area was covered by forests of deciduous trees. Conifers were generally small and found mostly in the higher areas, like they are today in the high mountain regions of the Western Hemisphere. Then, about one million years ago, before the ice age, the conifers probably invaded the entire region and became large, while the deciduous trees almost completely disappeared. Today, deciduous trees (like the maple and aspen) survive only in disturbed areas, where the conifers have not had enough time to reestablish their dominance.

A possible reason for this is the very peculiar climatic history of the region. During the last ten million years, as the coastal mountain chains and the Cascades were rising, they increasingly blocked clouds loaded with moisture coming from the ocean. Winters became moist and mild. Summers, although drier than winters, were made milder by the cold air currents flowing down from the high mountain peaks. This cold air would cause a drop in temperature, especially at night. The entire region was often covered by mist, which still happens today. The climate is mild, and frost is very rare, at least on the plains.

The Advantages of Conifers

Even in the mountains, the soil is very rarely frozen, due to the rapid covering of a thick layer of snow which acts as an insulator. In these conditions, the conifers are able to take water from the soil all year round, and to carry out

Opposite page: A large forest of conifers stands in the American Northwest. In the foreground, a river has been blocked by a beaver dam. The dominance of conifer trees, which are found in all forests of this region (not only in mountain forests), is found nowhere else in the world.

The multicolored clearing in the foreground overlooks a wide expanse of conifer forest on the Rocky Mountains. Conifers have ideal conditions for their growth and development in the cool and moist climate of the northwestern Pacific region.

photosynthesis even when the temperature drops below the freezing point. Photosynthesis, you may remember, is the process by which plants combine carbon dioxide with water to make food. It requires the presence of chlorophyll (a green pigment contained in leaves), that is, it requires leaves to be on the tree. For this reason, deciduous trees are not able to carry out photosynthesis during the winter because they lose their leaves. They are at a disadvantage when compared to the conifers, which can keep growing year round.

Conifers can store large amounts of water, minerals, and sugars (food) in their huge trunks. This can be an advantage during dry summers, when water becomes rare. Also, they can easily replace their reserves during cool summer nights. On the other hand, the trunks of the deciduous trees cannot store enough water to satisfy the plants' needs for twenty-four hours, because a lot of water is lost by transpiration from their flat leaves.

Also, deciduous trees usually store smaller amounts of sugars, though they need more than the conifers. In fact, conifers renew about 15 percent of their foliage each year, while deciduous trees renew 100 percent of their foliage. Also, conifers can recycle up to two-thirds of the food in their leaves before they shed them, while deciduous trees cannot recycle more than one-third. The food is recycled by being moved to other parts of the tree before the leaves are shed. Still another advantage results from the shape of the conifers. Their cone-shaped canopy (leaves and branches) allows them to catch the sun's rays very well, even when the sun is low on the horizon or the sky is cloudy. Deciduous trees usually have many of their leaves in the shade of others at any one time.

At higher altitudes, where snow is more common, there are some species of hemlock and fir trees. They are easy to recognize by their more slender shape, and their branches are bent downward. These trees manage to get rid of the snow, which slips down their branches, so they can carry on photosynthesis all winter. Also, their thin needles dry quickly, allowing some transpiration during the winter. This is necessary so that nutrients can be absorbed along with water from the ground. In summer, the needles collect tiny drops of water from the mist in the air, which gives the trees another supply of water, besides rain.

Finally, the conifers contain many protective chemical substances, like tannins and terpenes. These help make

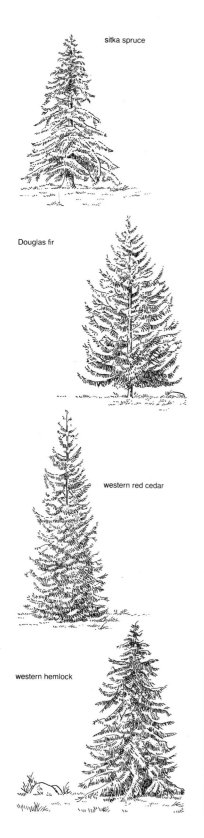

them resistant to diseases and insects. Obviously, conifers have found a very favorable environment in the cool and moist climate of the Pacific region. They have adapted perfectly to the local conditions, often reaching extremely large sizes.

Various Species and Their Natural History

Many species of conifers exist today. Each is different from the other. Some live only in certain habitats. The sitka spruce, for example, is usually found on the coastal plains and can rarely be found inland. Douglas firs are easily recognized by their cones. The cones are 3 to 4 inches (7.5 to 10 cm) long with three-pointed bracts or scales sticking out of them. Their needles, which are up to 1.5 inches (4 cm) long, stick out straight in all directions from the twigs. Western red cedar was often used by the Indians to make totem poles and canoes. Its wood is actually very durable, and today it is highly prized for use in floors and roofs. The leaves are scale-like, and the reddish bark tends to come off in strips. The cones are small and hang from low-lying branches.

The western hemlock has flat needles, only 0.5 inch (1.3 cm) long. Together with the sitka spruce, it is the most common conifer in the lowland forests of this region. It also reaches the western slopes of the Olympic Mountains and the Cascades. In western California, on the other hand, the largest of all conifers is the redwood. This giant can grow only on the plains and low hills and does not extend up the mountain slopes.

When the largest trees die and fall to the ground, all of the nutrients in them are recycled into other organisms. Very soon, an entire community of insects has consumed much of the fallen giant, helping its rapid breakdown. Over three hundred species of organisms live in the dead conifers. The metallic woodboring beetle for example, feeds on needles of living trees, but it lays its eggs in cracks in the bark or in crevices of recently fallen trees. The larvae will eat the sap and the still-green wood.

The relatively rapid rotting of the tree allows nutrients to be recycled or turned into humus (decomposed matter). On the dead trunks, then, new seedlings and small trees will grow. These use the trunk as a kind of nursery. Usually, only one of these trees will grow to maturity. Eventually it will fall and die, struck by a storm or by some disease, and the cycle starts over.

The giant sequoia and the redwood are certainly among the largest trees in the world. They can rise to 394 feet (120 m), the height of a forty-story building. They can be 32 feet (10 m) wide and reach the age of 2,200 years. These trees can reproduce by sprouts from their base. Due to their limited area of distribution and the cutting they had undergone, they were close to extinction. Luckily, today, they are protected inside national parks and reserves.

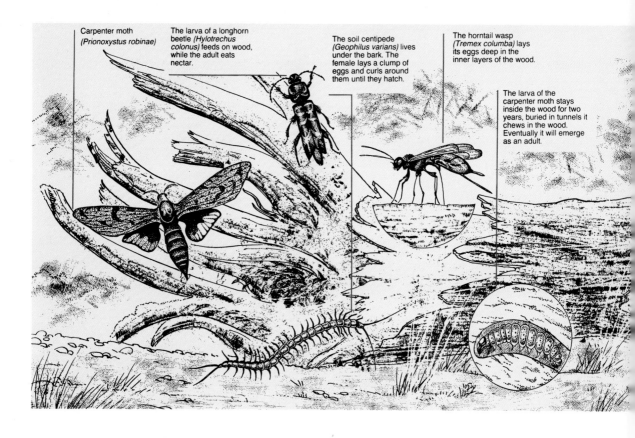

Carpenter moth (*Prionoxystus robinae*)

The larva of a longhorn beetle (*Hylotrechus colonus*) feeds on wood, while the adult eats nectar.

The soil centipede (*Geophilus varians*) lives under the bark. The female lays a clump of eggs and curls around them until they hatch.

The horntail wasp (*Tremex columba*) lays its eggs deep in the inner layers of the wood.

The larva of the carpenter moth stays inside the wood for two years, buried in tunnels it chews in the wood. Eventually it will emerge as an adult.

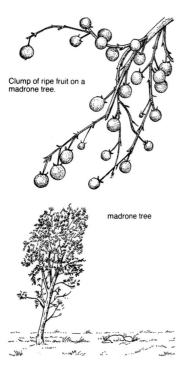

Clump of ripe fruit on a madrone tree.

madrone tree

Today, the Douglas fir tree is the dominant species in many secondary forests. Secondary forests are those which have regrown, having been cut down. These trees grow rather quickly, but they cannot tolerate shade. For this reason, they are the first to come back in open areas formed by tree cutting. Cedars and the hemlock, on the other hand, can grow in the shade, so they colonize forests where the Douglas fir is dominant. Eventually—often after hundreds of years—these species manage to gain more space and to become dominant. The climax forest is a forest in its final, balanced stage of development. It is composed mainly of the cedars and the hemlock. On the other hand, the Douglas fir is a typical "pioneer" species, as it can grow in environments which are not suited for other species. Very few Douglas fir will survive in a fully developed forest.

Deciduous Trees and the Underbrush

Besides the conifers, some species of deciduous trees grow in the forests of the moist plains. These trees, though, never cover wide areas, except for some recently deforested

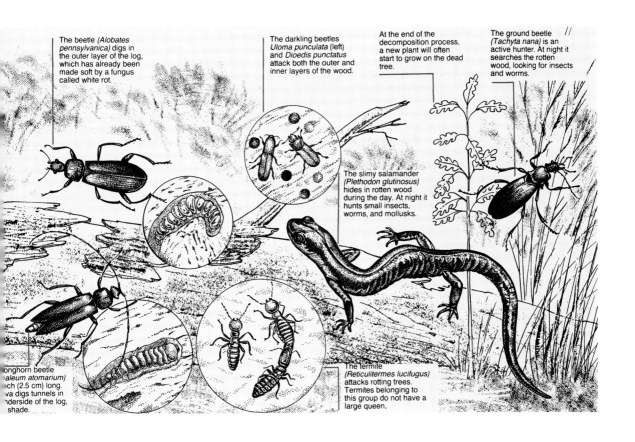

The beetle *(Alobates pennsylvanica)* digs in the outer layer of the log, which has already been made soft by a fungus called white rot.

The darkling beetles *Uloma punculata* (left) and *Dioedis punctatus* attack both the outer and inner layers of the wood.

At the end of the decomposition process, a new plant will often start to grow on the dead tree.

The ground beetle *(Tachyta nana)* is an active hunter. At night it searches the rotten wood, looking for insects and worms.

The slimy salamander *(Plethodon glutinosus)* hides in rotten wood during the day. At night it hunts small insects, worms, and mollusks.

The termite *(Reticulitermes lucifugus)* attacks rotting trees. Termites belonging to this group do not have a large queen.

onghorn beetle *aleum atomarium)* ch (2.5 cm) long. va digs tunnels in nderside of the log, shade.

Above: After growing for hundreds of years, the great trees die and fall to the ground. Here, through a slow process, the material they are made of returns to the soil. This decomposition is partly caused by weathering, but it is mainly the work of insects, fungi, and bacteria. In the drawing, some of the small or tiny organisms which can be found in the rotten trunks of American forests are at work.

areas. One common deciduous tree is the bigleaf maple. This tree can grow quite large, up to 3 feet (1 m) in diameter. Like the vine maple, this tree has large leaves which turn red and gold in autumn, adding a touch of color to forests which would otherwise be green. Another tree often found in the region is the Pacific dogwood. This species is especially attractive in spring, when it blossoms with large pinkish white flowers.

However, the most beautiful deciduous tree of this area is certainly the Pacific madrone. This tree does not shed its leaves. It has a closely related cousin in Europe. Its bark is reddish, and easily pulls away in long strips. In autumn, madrone trees produce large amounts of orange-red berries, which are excellent food for birds like the robin. The seeds go through the birds' stomach and intestine undigested and are eventually eliminated with the droppings, often far away from the tree which produced them. This helps spread the species to new areas.

The underbrush of the rain forest is often a dense tangle of shrubs, ferns, and rocks covered with moss. Among

the shrubs which should be mentioned is the "devil's club." This plant gets its name from the sharp and tough thorns it has all over its branches and leaves. The shrub can grow to be 10 feet (3 m) tall, and is a real nuisance to hikers who are not careful about where they walk.

The Fauna: Small Animals

The fauna in coniferous forests is perhaps not as varied and plentiful as in other environments, but it does include several interesting creatures. A particularly showy invertebrate is the banana slug. This is a large ground mollusk, which can be up to 6 inches (15 cm) long. It is usually an olive-yellow or brown color and is sometimes spotted. It lives in thick forests, feeding only on plant matter.

Among the vertebrates, amphibians like to live in the forest, which is always moist and with numerous pools and puddles of water. A very widespread species, which is more readily heard than seen, is the Pacific tree frog. In early spring, these animals gather in pools to lay their eggs, and their call (which has been described as "rivit" or "rikit") can

An incredible amount of mosses can live on trees in northwestern American forests. Among them, the most spectacular are the hanging mosses. These sometimes hang down from branches like a very long fringe, as seen here.

Pacific tree frog

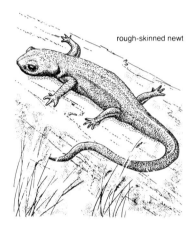

rough-skinned newt

be heard everywhere. The females lay small clusters of eggs on plants in shallow water. In May, the reproduction period is over, and the adults leave the pools. They will spend the rest of the year in the dense underbrush. Then the calls become less frequent, but they can still be heard throughout the year. Actually, winter on the moist plains is usually very mild, and the tree frogs can remain active all year. Another amphibian, the rough-skinned newt, reproduces in pools starting in late autumn or early winter.

Not many species of reptiles live in the region. Only one lizard, the northern alligator lizard, can be considered common. This little animal is some 10 inches (25 cm) long. It hides under rocks and logs, and mostly feeds on invertebrates. It is a viviparous species, which means it gives birth to its young rather than laying eggs. The female gives birth to two to fifteen young. They soon free themselves from the thin membranes in which they are wrapped when they are born.

The most widespread snake in the area is the northwestern garter snake. It can be as long as 24 inches (60 cm). It is non-poisonous. Its color is extremely variable. Some are striped, some an even color, some are olive or brown, some others have bright yellowish or red bands. These snakes live

Right: Slugs are land mollusks similar to snails, but unlike snails, they do not have shells. Some slugs, like the banana slug in the picture, can be quite large and brightly colored.

varied thrush

winter wren

dusky grouse

spotted owl

in different habitats, and they can be found in clearings or meadows, as well as in the forest. Like the alligator lizards, garter snakes are also viviparous.

Birds

In spring and summer, the song of the varied thrush is one of the most familiar sounds in the rain forest. It is a piercing song, sung in one note but repeated at different pitches. Sometimes it sounds like a whistle or a hum, then it turns into the jingle of a light silver bell.

Often, from the dense tangle of the underbrush, the long and complex song of a winter wren can be heard. From time to time, the forest is filled with the booming call of a male dusky grouse. This sound is so low that some people can hardly hear it. The booms and low "hoots" are produced by special sacs on the sides of the bird's neck. These sacs, which become filled with air, are a bright orange color and are fringed by white feathers. The booming call and the elaborate courtship dance which comes with it are meant to attract the females. After mating, the females lay their eggs, sit on them, and raise the young, without any help from the males.

On steep hillsides, the spotted owl can sometimes be found. This species likes fully grown conifer forests and drier locations. A lot of tree cutting has seriously harmed it. The frequent call of this owl can be heard in particular during the reproductive period. It can irritate hikers and campers, even those most knowledgeable about and interested in nature. The call sounds like a continuous sequence of coughs, barks, and high-pitched hoots. It can sometimes be tricked into echoing a person trying to imitate it. The spotted owl feeds mainly on rodents and birds, and builds its nest in the spring.

Another rather common dweller of the rain forest, found mainly in clearings and along riverbanks, is the song sparrow. Like many other birds in the area, this sparrow has rather dull coloring. It makes up for this by its full and varied song, from which it has gained its name. On clear spring mornings, few forest sounds are more delightful than that of a choir of song sparrows. They are warning their rivals to keep away from their territory, and at the same time they are inviting the females to join them. Despite the high humidity of the forest environment, the song sparrows almost always build their nests on the ground, at the bottom of a thick tangle of vegetation.

shrew mole

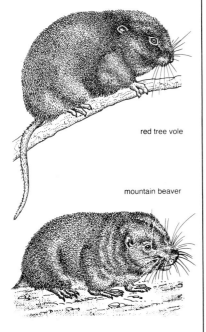
red tree vole

mountain beaver

black bear

Mammals

The most peculiar of the native mammals of the Pacific Northwest is probably the shrew mole. This animal is actually a mole, but many of its features are similar to a shrew's. For example, it has a furry tail and a rather slender body. Like all moles, the shrew mole has tiny eyes, covered by a layer of skin. It can probably distinguish light from darkness but not much more. Its sense of smell, on the other hand, is extremely developed. Probably the shrew mole uses this to search for its food. These moles also differ from other moles in that they often venture out of their tunnels and hunt insects and other invertebrates on the surface. Usually their tunnels are not very deep. Sometimes they are merely passages just below the layer of dead leaves. The reproductive period of these animals is rather long. The young are usually born between February and November.

Another interesting native mammal is the red tree vole. It is similar to the other voles, but it has a long tail, covered by blackish hairs. The color of the tail contrasts with the vole's bright reddish body. But the most peculiar thing about this species is that, unlike its relatives, the red tree vole does not live on the ground. It is strictly a tree dweller. It lives mainly in the conifer forests of coastal Oregon and northern California, and only eats conifer needles. Like many other tree rodents, the red tree vole builds a large nest with twigs and leaves, and attaches it to the branches nearest to the tree trunk. The nest grows bigger and bigger every year, and sometimes is used by different generations of voles. The young can be born any time throughout the year, as is the case with the shrew moles.

A primitive species of rodent, also native to the coastal area, is the mountain beaver. This mammal digs long tunnels under tangled thickets, in the underbrush of conifer forests, or in narrow clearings. It actually does not resemble a beaver, either in its look or in its habits. It is not a true beaver and is classified in a different family of rodents. It measures 12 to 18 inches (30 to 47 cm). It has a very short tail, small rounded ears, and nocturnal habits. Although it is often found close to the water, it prefers dry land and is strictly a vegetarian.

Among the large forest mammals, the black bear is one of the most shy. Males weigh between 220 and 550 pounds (100 to 250 kg), females between 90 and 175 pounds (40 to 80 kg). This bear usually prefers to be alone, and roams about within its territory. A territory is often between 2 and 6

Two black bear cubs, perhaps frightened by the photographer, hurry to shelter by climbing two trees. The black bear is the smallest, most common, and best-known species of its family in the United States. It can easily be observed, even at close range, in many national parks and wildlife refuges. The cubs stay with their mother for two years, until they learn to gather their own food. Bears are omnivorous, which means they eat various foods, such as roots, grass, fruit, insects, rodents, and carrion.

A mule deer doe quickly crosses a marshy area in Yellowstone National Park. The mule deer is more primitive than the European red deer, but the two species are similar in many ways, both in their physical features and in their habits.

sq. miles (5 to 16 sq. km) for the female, and 20 to 27 sq. miles (50 to 70 sq. km) for the male. In autumn, just before going to their dens or caves to hibernate, bears build up a large amount of fat. This will allow them to survive during the many months of hibernation. During the winter, though, especially if the weather stays rather mild, bears can come out of their refuge at certain times. In fact, their body temperature does not greatly decrease, and they easily wake up from their hibernation. If, on the other hand, the winter is cold, they will hibernate for long periods, without touching food or water.

The females mate when they are two to three years old, and then every other year for their entire lives. They can live

buck mule deer

to be ten to twelve years old. The cubs are born inside the winter den, usually two, sometimes three of them. At birth they are naked, their eyes are closed, and they weigh only 0.5 to 1.5 pounds (230 to 700 g). They soon start feeding on their mother's milk.

The most common hoofed mammal of the plains forests is the mule deer. This deer is more primitive than deer of Europe, Africa, and Asia. It has shorter teeth and larger bones and toes, which allow it to move easily on soft or sinking ground. On the other hand, this animal is also well adapted to moving on steep slopes, rocky areas, and even in deserts. It is found all over the region, from the ocean to the Rocky Mountains.

MOUNTAIN FORESTS, ALPINE MEADOWS, AND PEAKS

Moving inland from the coast, the mountain peaks become much higher, until reaching the Olympic Mountain Range and the Cascades. Around 3,281 feet (1,000 m) in altitude, the forest begins to change its appearance, and at 6,562 feet (2,000 m) it has a completely different look.

Trees and Flowers

The main trees are still conifers in this area, but these conifers are of different species. The Englemann spruce is one of the most common. It is easily recognized by its tall, even shape. Its bark is scaly, purplish to reddish brown. Also common, especially at higher altitudes, are mountain hemlocks and subalpine firs. Wide stretches of lodgepole pine may sometimes be a thick barrier in sandy and moister areas. This pine owes its name to the fact that the Indians of these regions used its long, straight trunks as lodgepoles for their dwellings.

The conifers just described also dominate the Rocky Mountains. At higher elevations, larches, which are conifers that shed their needles, can also be found. In autumn, before falling, their needles turn reddish or golden, and add a bright touch of color to the gray and hazy mountain background.

Higher still, and especially on the peaks, subalpine firs and mountain hemlocks take on weird, twisted shapes due to the continuous winds. The winds often blow snow and ice onto the trees. For this reason, the vegetation grows in odd shapes, sometimes even sideways, close to the ground, not higher than 1.5 feet (0.5 m). The short and knotty trees found here are called "elfinwood." It is hard to believe that they belong to the same species as the tall and swaying trees which grow at lower altitudes and in more sheltered habitats.

Another strange thing in higher altitudes is the presence of long rows of trees on rounded ridges and on plateaus. These formations are called "ribbon-forests." Scientists still do not have a good explanation of how they form. Perhaps winter snowstorms play some part in the process. In any case, once the ribbon-forests are formed, they act as snow shields. Snowbanks several feet high collect around them, creating a protective barrier.

Also, some deciduous trees live in the region, especially in sheltered valleys or in areas which have been burned by fire or hit by avalanches. The most widely distributed and well known is the quaking aspen, found all

Opposite page: Mule deer graze in a conifer forest at high altitude. At these altitudes, the main conifer species are Englemann spruce, the mountain hemlock and the subalpine fir. Higher up, larches, which are deciduous conifers, are also found.

Right: A species of flower belonging to the Compositae family blooms on a northwestern meadow. On the alpine meadows of the Rocky Mountains, as well as on those of the European mountains, "spring," which is the blooming period, is often delayed until July or August. Many flowers have evolved mechanisms meant to keep the temperature inside them much higher than the environmental temperature.

Opposite page: The sap from stems and flowers of *Ranunculus acris*, a buttercup of European origin, is very acidic, and keeps herbivores from eating it. For this reason this plant is common all over the Rocky Mountains. Locally called "elephant's head," due to the shape of its flowers, *Pedicularis groenlandica* grows in wet grasslands at medium altitudes in northern California. *Penstemon whippleanus,* on the other hand, grows on alpine meadows and gentle rocky slopes at timberline. It is called "beardtongue" and grows in Montana, Idaho, Wyoming, and all the way to Utah and northern Arizona. *Castilleja miniata,* or Indian paintbrush, is frequently found in northwestern American mountains, from Alaska to California.

Ranunculus acris

Pedicularis groenlandica

Penstemon whippleanus

Castilleja miniata

over in the area. This tree has a smooth, olive green to cream-colored bark, with darker, warty patches.

Often spring in the alpine meadows does not come until July or August. But when it comes, it brings wonders. Brightly colored flowers completely cover the ground. Several species are especially adapted to life at high elevation, where the days are cool and nights are cold.

Inside some flowers, buttercups, for example, temperatures can be 40° to 41°F (4° to 5°C) higher than the air temperature, especially when the sun shines. During the day, other flowers bend and follow the path of the sun. This property is called "heliotropism." Others, like the anemones, cannot move with the sun, but are still able to catch the daytime heat. Their petals reflect the sun's rays inward, and the flower warms up. The outer parts of these flowers, called "sepals," protect them from the cold winds. In some species, the inside temperature has been recorded to be up to 57° to 64°F (14° to 18°C) higher than the air temperature. The accumulated heat speeds up plant growth, especially in northern or mountain environments. Here the climate is most extreme, and even in the middle of summer the air temperature is only a few degrees above the freezing point. Through the use of heat-collecting methods, the number of days fit for growth can be increased up to 25 percent. The production of pollen and seeds is also increased, thus attracting more pollinating insects.

Many insects which take advantage of the locally increased heat are often found on these flowers. These small animals have no means of regulating their body temperature. When inside the flowers, their temperature can be up to 90°F (32°C) above air temperature. Some insects even spend the night inside flowers, or rest in them during the day, thus reducing their energy needs. Otherwise they would need to use extra energy to keep warm. The plants in turn also benefit from this insect habit because the tiny creatures carry their pollen from flower to flower, aiding in pollination.

Mammals

Walking through the mountain forest, all is often quiet, and there seems to be no sign of animal life, even in the middle of summer. This first impression, though, is completely wrong. Actually these forests are home for a surprising number of animals, although not as many as are found in deciduous or mixed woods.

The wapiti is a close relative of the European red deer, but it is larger and lives more to the north. In the United States, it is called "elk," a name that in Europe is used to indicate a moose.

A characteristic of the conifer forests is the deep silence, broken from time to time by sudden bursts of activity. A person may hike for a long time without seeing or hearing a thing, and then suddenly come across a group of deer. Or, a visitor may get close to the perch of a songbird, which will sing at the top of its voice to signal its territory. Only in springtime does the sound of singing birds become common.

Various species of deer live in the mountain forests. The mule deer and its closest relative, the white-tailed deer, can be found all over the western region. The white-tailed deer stays more confined to forests and swamps. It rarely comes out in the open, like the mule deer does. The white-tailed deer browses mostly on twigs and shrubs, but during the summer it also eats grass. Like the mule deer, it mates in autumn, and the fawns, usually one or two, are born in spring or early summer.

The third species frequently found in the region, the largest of all, is the wapiti or elk, a close relative of the European red deer. This is a large-hoofed mammal, able to

The moose is the largest species of the Cervidae (deer) family and also the largest of all animals living in the American Northwest. Meeting a large male moose close up is nearly as memorable as meeting an elephant. In the picture, three males show off their typical palm-like antlers.

feed on vegetation up to 9 feet (3 m) above the ground. It is easy to recognize, not only by its size, but also because it has a pale yellowish patch on its back. During the summer, it is found mainly in the open forests and in meadows bordering the woods. In winter, on the other hand, it moves down to the bottoms of the valleys, near the foothills. During the mating season, the bucks develop huge antlers and give off loud bellows which can be heard several miles away. The strongest bucks will manage to gather a large harem and will have many battles with their rivals in order to keep control of the does. The fawns are born in May and June.

By far the largest of the hoofed mammals belonging to the Cervidae (deer) family is the moose. It can be 6 feet tall (2 m) at the shoulder, and it lives almost entirely in thick forests and near swamps. The moose is most likely to be found in the mountains of Montana, Idaho, and Wyoming, and also in Canada. It has rather poor eyesight, but very good hearing and sense of smell. This mammal has a large body on rather long, thin legs. Its hoofs are widened at the bottom. This allows the moose to walk on fresh snow and across swamps and marshes without sinking.

A mainly nocturnal animal, the moose browses on water plants during the summer, and on twigs and shrubs

Below: The drawings show one of the spectacular duels which the bighorns engage in each autumn. When the mutual display of large horns is not enough to discourage one of the two males, the animals charge head-on with great violence.

Opposite page: The horns of male bighorns do not reach full size before the animal is six years of age. Thus it is very hard for young males to have and control a harem. The drawings show some stages of the horns' growth. The tips of the horns of the older animals are curved to form a complete circle.

during the winter. The moose has a large "nose" and a dewlap, which is loose skin that hangs from the neck. During the mating season, bulls develop extremely large antlers, which are different from those of all other species because they are palm-shaped.

Higher up, on rocky ridges and boulders, and especially in the Rocky Mountains, one can find the well-known bighorn sheep. It is easy to identify; its fur is brown with a lighter back, and the rams (males) have large, curved horns. The females have much smaller horns, which do not curve up into a spiral.

Bighorns are skilled climbers, able to run at high speed up steep—almost vertical—cliffs. Their hoofs have sharp edges and are concave. This means the hooves are hollowed out in the center. This allows them to grip onto the smallest projections of the rock and climb almost vertical slopes. These animals, like many other hoofed mammals, live only in separated groups of either rams, or females and young. They graze among the hard-to-reach rocks and on the high-altitude meadows. They usually do not enter into the forest.

Living in open spaces, the bighorn has extremely sharp eyesight, unlike the moose. The moose lives in thick forests, so it relies more on its senses of smell and hearing. Those two senses, on the other hand, are not very well developed in the bighorn.

Another animal of the ridges and peaks of the highest mountains, the mountain goat, lives even higher up than the bighorn. Actually, this species is not a true goat. It is a relative of the European chamois. This animal can be found on the highest peaks of the Rocky Mountains, the Cascades, and the Olympic Mountains.

growth at two years

growth at six years

growth at twelve years

The mountain goat is very well adapted to life at high elevations. In winter it develops a thick white fur. Its hoofs are equipped with a convex cartilage "cushion" with a hard edge. This provides an anti-skid surface. The mountain goat is perhaps an even more spectacular climber than the bighorn. It can climb up almost vertical surfaces and perform leaps of several feet from one tiny rock projection to the next. The young, born in May and June, are well developed. They are able to follow their mothers up the most difficult trails a few minutes after their birth. The small, curved, and

Snow-white mountain goats are typical dwellers of the mountains of British Columbia and a few other regions. They are rather closely related to the European chamois. They like the higher slopes and are skilled jumpers and climbers.

very sharp horns of the mountain goat resemble those of the chamois and are used in frequent fights, especially among males.

A strange feature of this animal is the rump shield. This is a patch of thickened skin around the groin and on the rump. The rump shield is probably a protection against possible injuries that could be caused by a rival's sharp horns during fights. Despite this, many adults have at least some scars, probably the result of the frequent fights.

During the summer, mountain goats graze at high altitudes. In winter, on the other hand, they move downhill to avoid snowstorms. Food is not easy to find during this season, so the animals gradually use body fat that they built up during the summer.

The pika is another animal that lives in the highest forests and at the timberline. This strange animal, which resembles a rodent, is usually found on large piles of

boulders and rubble. A gray-brown color, this creature has short, rounded ears and practically no tail. It is not a real rodent, but instead it is related to rabbits and hares, which all belong to the order Lagomorpha. Pikas feed only on plant matter, which they gather in the meadows and forests. Like all other lagomorphs, they produce two different kinds of droppings. One is a kind of jelly-like fecal pellet, which is eaten again and goes through the digestive tract one more time. This allows for a very efficient absorption of food nutrients.

Usually pikas live at altitudes over 6,562 feet (2,000 m). They seek shelter from the strong summer sun and from winter snowstorms among the rocks. There they also find a safe shelter from most predators. Weasels are an exception, however. They can reach pikas even through the narrowest cracks and tunnels. The pika's territorial call, made frequently even when a predator is in sight, is a short, sharp whistle. This sound is familiar to all hikers on western American mountains.

Pikas mate in the spring, and in summer one to two young are born. When the young are old enough, the food-gathering period starts. Each individual gathers various plant material from the meadows and stores it in its underground burrow. Unlike the mountain rodents, pikas do not hibernate. They stay active throughout the winter. Their thick fur is good protection against low temperatures. If stored food is not enough to last through the winter, pikas can gather some more. They dig tunnels in the snow until they reach some buried grass. From time to time, these tiny animals will even come out in the open in the middle of winter, provided the weather is nice, with no blowing wind or storms.

Higher still, usually at or above the timberline, hoary marmots can be found. Eastward in the Rocky Mountains, these are replaced by the yellow-bellied marmot. Hoary marmots are large rodents, weighing up to 20 pounds (9 kg) and over 24 inches (60 cm) long. They are the largest species of the Sciuridae family, which also includes squirrels and chipmunks. Unlike pikas, these animals hibernate during the winter, and their body temperature can then drop to 39°F (4°C). By mid-May they come out of their complex system of tunnels to feed on plant matter on the alpine meadows.

Often marmots are on guard perched on a large boulder, ready to give out a warning in case of danger. For

Right: The small golden-mantled ground squirrel is one of the most common and best-known rodents of western North American forests and parks. Similar in many ways to the more common chipmunks, this is one of the typical hibernating mammals.

Below: Beavers have large incisor teeth which stick out from their closed mouths. With these teeth, they can even gnaw underwater. They are an important working tool for the beaver and are especially useful for cutting down trees and branches.

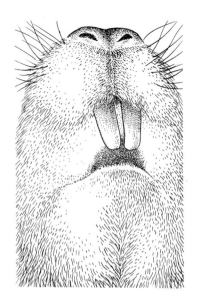

the hoary marmot, this warning is a loud whistle; the yellow-bellied marmot uses a series of high-pitched twitterings. In spring and summer, these are among the most familiar sounds in the mountains.

Usually, marmots go into their burrows in September or October, after having stored a large amount of fat which they will use during hibernation. Some marmots, which live in very cold locations, can hibernate up to eight months a year.

Another hibernating mammal, living in the high mountain forests and on the alpine meadows, is the small golden-mantled ground squirrel. It is in many ways similar to a chipmunk, but it is larger and does not have stripes on its face. Instead it has a cream-colored stripe with black edges on its sides. This squirrel is omnivorous, which means it eats

Below: This drawing shows beaver footprints on a muddy surface. The small prints of the animal's front paws, the large prints of its webbed hind paws, and the wide stripe made by it flat tail can all be clearly seen. Besides footprints, droppings (to the right in the drawing) also tell of the beaver's presence.

Bottom: A beaver swims in a stream in the Vancouver area of British Columbia. Beavers feed mainly on leaves and twigs, and they use logs and branches only as building material.

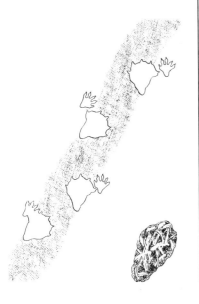

both plants and animals. It feeds on seeds, plant matter, insects, and even on carrion. It digs its burrow under bushes, logs or boulders, and goes into it in September to hibernate. But, like many other animals which hibernate to survive the winter, it wakes up from time to time. In early spring it will even come out into the open, digging tunnels through the snow.

The beaver lives at lower elevations in deep valleys and on alluvial (river valley) plains. This large rodent is up to 3 feet (106 cm) long and weighs some 60 pounds (27 kg). It is easily distinguished from all other similar species by its size and also by its wide, flat tail, which is shaped like a paddle. Beavers live near streams and ponds, and they are often noticed by the sound of their tails slapping on the water. To the beavers, this is the alarm sound, to which the entire colony always reacts by diving into the water.

Beavers are skilled builders. They use their strong front teeth to cut down trees, the branches and logs of which they use to build lodges and dams. Using their large incisors (gnawing teeth) and powerful chewing muscles, beavers can cut down trees 8 inches (20 cm) in diameter or more. They start by gnawing a ring of bark at the bottom of the tree. Then they go on, working deeper and deeper into the wood. Branches gnawed with cone-shaped ends are a clear sign of the beaver's presence. The cut trees, with their leaves and twigs, will provide a great deal of food. The inedible parts, such as the trunk and large branches, will be used as building material. With logs, branches, and other materials,

Unlike the European porcupine, the American porcupine is not a ground animal. Instead it is a tree dweller. It is covered by long hairs which hide its quills. This often causes its enemies to underestimate how dangerous it is. The porcupine can cause serious damage to the vegetation. It feeds in trees, on bark and leaves, and also on grasses and wild fruit. In relation to its adult size, the newborn porcupines are much larger than any other mammal's young.

the entire beaver family joins in building a dam or lodge. They cement the structure with clay, which dries in the summer sun or freezes during the winter. In both cases, the result is solid construction, a good defense against predators.

The water is used to transport logs and branches which will strengthen the dam or lodge. Beavers will carry some of the branches to the deep water. There they will be saved along the muddy bottom until winter. Then they will be used for food. Because the beavers do not hibernate, they have a hard time finding fresh food during the winter. Even

when the temperature gets very low, beavers can enter the frozen pond through an underwater passage connected to the lodge. They can then reach their food supply, safely stored under a layer of ice. Some beavers, living in large lakes or swift rivers, may dig dens in the banks. These are usually dug within or close to the thick wall of the dam, so they will be well sheltered against the cold.

Another important rodent, found both in plains and mountain forests, is the porcupine. It can weigh up to 22 pounds (10 kg). It is mainly nocturnal and lives alone. When on the ground, this animal looks clumsy and awkward. It moves about in a cumbersome way with short steps. When in trees, on the other hand, it moves easily, feeding on bark, small twigs, and leaves. Like many other wild mammals, it eagerly seeks salt. There is a good chance of seeing it where salt is left for deer, or walking along the roads, where salt is thrown to melt snow and ice in the winter.

The most outstanding features of this species are the long, sharp quills which cover its back. When threatened by an enemy, the porcupine turns its back, raises its quills, and lashes its tail, which is also well armed with quills. If a predator attacks, many quills will come loose and stick into the enemy's skin. Pulling out these "arrows" is a difficult and painful operation, due to the tiny barbs pointing backwards which cover the ends of the quills. If the quills are not promptly removed, they can work their way deeper and deeper into the body. A predator with quills stuck in its face or mouth may not be able to eat, and will actually starve to death. For this reason, most predators carefully avoid porcupines. Some, though, like the wolverine, can kill them and eat them. Probably they attack their victims by surprise, roll them on their backs and bite into their belly, which has no quills.

Predators

The largest and most frightening carnivore living in this region is certainly the grizzly bear. This animal was once rather common in all the western mountains and plains, but it has been killed off almost everywhere. Today it only survives in certain protected mountain areas, like Yellowstone National Park, Glacier National Park, and some small, isolated areas in Montana, Idaho, and Wyoming. In the Canadian Rocky Mountains, though, it is still found rather frequently.

Grizzly bears are huge animals, sometimes standing as

Three grizzly bears fish for salmon, while a small group of gulls waits patiently for some leftovers. Grizzly bears are the largest ground carnivores in the world. Large males are over 9 feet (3 m) long and 4 feet (1.2 m) tall at the shoulder. Bear populations have sharply decreased since the slaughter of buffalo. In fact, the buffalo were an important part of their diet.

Right: A mountain lion tries to catch a raccoon (seen from the rear) in a tree. This predator is quite a large animal, larger than a leopard. Its size justifies its name, *mountain lion.*

Opposite page: The bobcat is seen in several different hunting postures. This is another typical North American predator belonging to the Felidae family. The bobcat is a little smaller than the lynx.

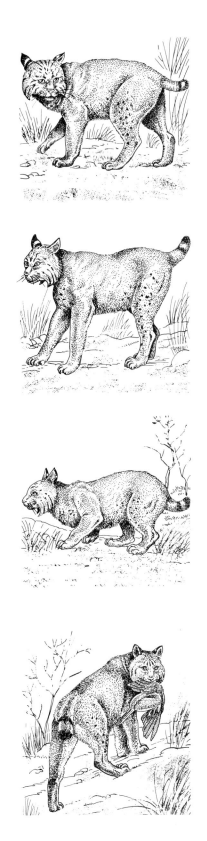

tall as 8 feet (2.5 m) and weighing 1,543 pounds (700 kg). Besides their size, grizzly bears are easily distinguished from the smaller black bears by their clearly humped backs and flat faces. Except for when the females are with their cubs, these bears live alone. They roam the open forests, the plateaus, and alpine meadows above the timberline, gathering fruit, leaves, insects, and larvae. They use their powerful front paws, armed with long claws, to dig for rodents, like ground squirrels and marmots.

Grizzly bears are extremely powerful, and can run short distances at high speed. They have been recorded moving as fast as 31 miles (50 km) per hour. At times they can kill large game, like an elk or moose, but usually their victims are sick or somehow weakened animals. Also, grizzlies often visit streams during salmon migrations and are especially skilled fishers. Salmon provide a rich food supply in late summer and early autumn, just before the bears hibernate.

Grizzly bears, like black bears, are not truly hibernating animals because during hibernation, their body temperature does not drop very much, compared to ground squirrels and marmots. Bears may fall asleep for long periods in a den hidden under a large fallen tree, in a cave, or in some other shelter. But they can easily wake up and wander out into the open during the milder winter days. Two or three cubs are born in January or February. They are almost hairless, with their eyes closed, and are nursed and kept warm by their mother in the winter den.

The mountain lion is another big predator. This predator, however, is very difficult to see. Powerful and very nimble, this animal can be 6 feet (2 m) long and weigh 220 pounds (100 kg). It lives in mountain forests and also at lower elevations, in dry, wild areas. It is still common in some regions (in the Olympic Mountains, for example), but even there it is rarely seen. Most hikers, even those who have spent weeks or months hiking all over areas where the mountain lion lives, have never come across a mountain lion, even though its tracks or its remains can sometimes be seen.

The mountain lion is mainly a nocturnal animal and lives alone. It mostly hunts deer, rodents, and rabbits. A single animal can kill and eat up to fifty deer in a year. This predator is thus an important part of the ecosystem, which includes all plant and animal species living in an environment. For one thing, the mountain lion helps control deer

Above: A bobcat has just caught a rabbit. The cautious predator will probably carry its catch to a safe place before eating it.

Opposite page, from left to right: When two wolves of the same rank meet, they first stare at each other with their tails raised. Then they come closer and curl their lips, show their teeth, and flatten their ears. Finally, they fight each other standing on their hind legs and grabbing each other with their front paws. At the end of the fight, the loser rolls on its side and offers its throat to the winner.

population. It can help prevent overpopulation, over-grazing, and the eventual death by starvation of whole herds of deer. In fact, in the areas where the mountain lions have been wiped out, deer have increased their numbers too much, and eventually need to be killed by humans.

Two other species of the cat family live in the region: the lynx to the north, and the bobcat more to the south, in the coastal regions of Washington, Oregon, and California, and eastward. The lynx mostly lives in mountain forests. The bobcat, however, can be found all over, except for open plains, although it also prefers woods and canyons. Both species feed on rodents, rabbits, hares, birds, reptiles, insects, and berries. They can even kill young deer or antelope. Actually, they are quite large. The lynx is 3 feet (1 m) long and weighs 35 pounds (16 kg), while the bobcat is 2.5

feet long (76 cm) and weighs 24 to 30 pounds (11 to 13.5 kg).

The lynx populations rise and fall in relation to their favorite prey, the snowshoe hare. Roughly every eleven years, the hare populations increase to a high point and then usually suddenly fall to very low levels. During this period, the hares can be quite rare. The lynx largely reproduce when the hares are abundant. In times of famine, however, the predators starve to death, so their populations also fall.

Another large predator in this region is the wolf, belonging to the same species found in Europe. This animal, like the grizzly bear, has been hunted by humans. Today it is found only in hard-to-reach mountain areas, mostly in Canada. Unlike bears and the cats, this is a highly sociable species. Wolves live and hunt in packs, preying on large animals like deer and, in the past, also buffalo.

Hunting as a pack, wolves are very successful at killing their prey. While looking for prey, several pairs of eyes, nostrils, and ears are better than only one. When they need to catch prey, they can act together, with each member of the pack using its energy most efficiently. In summer, wolves also catch rodents, rabbits, and hares, and in autumn they like to eat wild fruit.

The wolf is the largest species of the dog family living in North America, Europe, and Asia. From the tip of the tail to the tip of its nose it can measure 6 feet (2 m), and it can weigh up to 115 pounds (52 kg). The color of its fur can range from black to gray to almost pure white. The pups are born

Below: The hummingbird can fly vertically and backwards, or hover in the air, flapping its wings. These skills are possible because of a special wing structure. The wing, which is very stiff, is connected to the shoulder by a rotating joint. It can make small movements at the "elbow" and "wrist." During the forward stroke (upper pictures), the wing moves to hold up the bird, without pushing it forward. During the backward stroke (lower pictures), the same result is achieved through a rotation of the wing of almost 180 degrees at the shoulder.

Opposite page: This drawing shows some of the typical birds of the northwestern American mountains. Hutton's vireo lives in moist forests. The purple finch lives in conifer forests, in canyon oak forests, and on lower mountain slopes. The red-breasted nuthatch lives in subalpine and western coniferous forests. The mountain bluebird is found in open meadows and grasslands, usually over 4,921 feet (1,500 m) in elevation. Clark's nutcracker lives on the higher mountain ridges, at the timberline.

in the spring and, after the nursing period, they are cared for by both parents, who stay together all their lives. In between hunts, wolves can go without food for several days. But when they kill prey they may stuff from 24 to 30 pounds (11 to 13.5 kg) of meat into their stomachs.

Birds

Over two hundred species of birds nest in the region, or regularly visit it during their migrations. In the conifer forests there are many species belonging to the Parulidae (wood warbler) family. This family is common in America, and plays a role somewhat similar to that of the Sylviidae family in Europe and Asia, whose members are known as old-world warblers and include many songbirds.

Often, in the spring, small wood warblers can be heard singing, together with Hutton's vireos. Western tanagers and the black-headed grosbeak can be found in mixed forests. Among other typical and common birds are the ruby-crowned kinglet, purple finch, hermit thrush, and woodpeckers such as the yellow-shafted flicker and the hairy woodpecker, which pecks on logs and branches of dead trees. In open areas, where flowers abound, there are hummingbirds. The rufous hummingbird and the Calliope

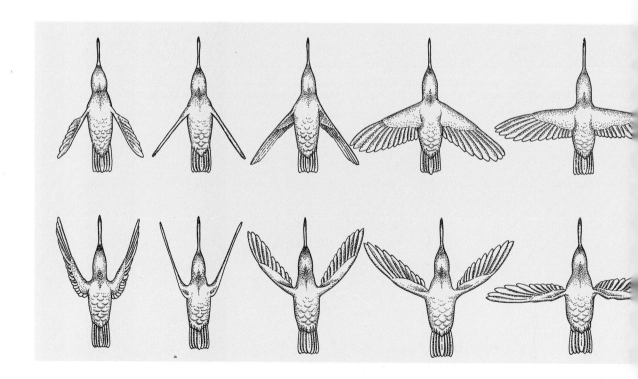

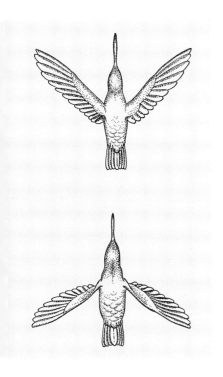

hummingbird are common all over in the mountain regions, while the black-chinned hummingbird and the broad-tailed hummingbird are more common in the eastern Rocky Mountains. These tiny birds are easily mistaken for large bumblebees as they fly from flower to flower. Some, like the Calliope hummingbird, are barely larger than some of the insects which live, as these birds do, on the flower's nectar.

All hummingbirds perform very complex and unusual courtship flights. The male rufous hummingbird, for example, flies in a wide, vertical circle, rising 98 to 131 feet (30 to 40 m) up, then swiftly nose-diving down toward the ground with a loud hissing sound.

Climbing higher, the number of bird species is lower, but there are some interesting and typical ones. In the most open areas and alpine meadows one may come across the mountain bluebird. The male of this species is such a bright blue that it looks almost unreal. Oddly enough, this color is not due to a pigment in its feathers, but instead it is caused by a peculiar structure of the feathers. This structure allows only the blue part of the light spectrum to be reflected. The

The Canada jay is a particularly friendly bird. It often comes near human dwellings looking for food.

females are more brown, but they too have some bright blue feathers, especially on their wings and tail. These birds nest in tree holes, sometimes in rock cracks, and will also use nesting boxes made by people.

Two other birds which are common at the timberline and above are the Clark's nutcracker and the gray jay, both belonging to the Corvidae family. These two species are rather common close to tourist facilities and often become very friendly. They come close to the picnic tables, looking for food. In fact, these birds are omnivorous, and will eat anything that is offered them.

Some interesting species live well above the timberline in the summer. Rosy finches, for example, gather in flocks on snowfields and glacier ridges. They nest in rock cracks at high elevations and feed mainly on seeds of alpine meadow plants and on insects, which can be abundant on the snowfields. The warm winds rising from the plains, in fact, carry many insects up to the mountains. As the winds cool off, the tiny animals are dropped on the snow. Rosy finches are not the only birds to take advantage of this rich food source. Other mountain birds, like juncos, sparrows, bluebirds, and pipits do, too.

One of the most typical mountain birds is the white-tailed ptarmigan, found above timberline both in the Rocky Mountains and in the Cascades. More to the north, in the Canadian mountains, two other species can be found. These species, which are present also in Europe and Asia, are the rock ptarmigan and willow ptarmigan.

Ptarmigans are known for their plumage (feathers) which keep them camouflaged all year round. During the summer, ptarmigans have drab colors that camouflage them. If they remain still among the rocks, they are very difficult to see. In winter they turn almost completely white, so once more they are almost invisible.

The plumage of these birds is also very thick, and it helps to protect them from the cold. When the climate becomes extreme, though, ptarmigans can dig holes and seek shelter in snowbanks. Their legs are completely feathered, which not only acts as insulation, but also allows these birds to move as if wearing snowshoes. They mainly feed on willow leaves and, in winter, also on conifer needles.

In winter, many birds leave the mountain forests and migrate south or downhill into the nearby plains. Some species, though, stay.

Frogs

Several species of frogs dwell on the mountain slopes, and some can live at altitudes of over 9,842 feet (3,000 m). The Cascades frog is found in lakes, pools, and quiet streams among the meadows, up to the timberline. It has a brown back with black spots, and a yellow underside. In spring, as soon as the snow has melted in the pools, the mating season begins. Then the males start their continuous "choir" of grating and chuckling sounds.

Another species often found in the pools and marshes along the mountain slopes is the Pacific tree frog. However, the oddest of all mountain amphibians is the tailed frog. This unusual animal is only found in the mountains and hills of northern California, Oregon, Washington, southwestern British Columbia, Idaho, and Western Montana. It lives at altitudes of up to 6,562 feet (2,000 m), both in mountain streams and in nearly stagnant pools. Its unusual feature is the male's reproductive organ which is so shaped that it resembles a tail. This organ is used to insure insemination (sperm transfer) even in the tumbling waters of mountain streams.

The tailed frog lives in mountain streams in the western United States (from Washington to northern California, Idaho, Montana, and Wyoming) and western Canada (southwestern British Columbia). The tadpoles of this frog have large sucker-like mouths. They can stand the force of the current and cling to the rocks and sometimes even to the legs of swimmers.

THE PLATEAUS

Wide stretches of grassland appear among the high peaks of the Rocky Mountains. These are the "seas of grass," which gradually slope down eastward and eventually join with the plateaus of the central part of the continent.

Huge Herds of Hoofed Mammals

Deep canyons, whose rims are covered with junipers and ponderosa pines, cut through the plateaus in every direction. Often, these areas are completely surrounded by high mountains, as, for example, in Yellowstone National Park.

Once, the grass of the grassland provided food for huge herds of hoofed mammals, like buffalo, pronghorns, and deer. Grasses of the Gramineae family, unlike other plants which grow from the top, do not stop growing after they have been cut or eaten. Instead, they produce new shoots from their root system.

Huge herds of buffalo were almost completely wiped out about a century ago. Today these animals are completely protected, and they have grown to thousands of animals. They are mostly found, though, inside wildlife refuges and national parks. Some are raised by private farmers and ranchers.

The males of this huge species can be up to 6 feet (2 m) tall and weigh 3,000 pounds (1,360 kg). These animals are perfectly adapted to the harsh climate of the plateaus. During the hot summer, they can go for days without a drop of water. During the coldest winters, they can sweep snow away with their heads and hoofs to get to the buried grass. During the harshest snowstorms they can stand against the wind, thanks to their thick, woolly fur, which acts like a shield. Usually though, during the winter, the herds move down to lower areas to avoid the extreme climate, and they may even make their way into the open forest.

Adult buffalo have very few enemies. Only grizzly bears and wolves dare to attack them, but even these powerful predators probably only kill old or sick animals. The young, on the other hand, are much more vulnerable, and their only defense is running. A few hours after they are born, they can already run with the herd, within which they find safety and protection.

Another hoofed mammal, the pronghorn, is found in grasslands, especially where the grass is not very tall, and also in sagebrush plains. This animal is quite unusual, since it is the only species left in the Antilocapridae family. It is

Opposite page: A male buffalo can be 6 feet (2 m) tall and weigh 3,000 pounds (1,360 kg). This animal is perfectly adapted to the harsh climate of the plateaus. Its thick, wooly fur allows it to stand wind and frost, even during the most violent snowstorms. Only the grizzly bear and the wolf dare attack it, but they do not always succeed.

Above: A male buffalo engages in duels at any time of the year, both to keep its social status and to attract females.
From left to right: After having warned each other with bellows, and having scratched the ground with their hoofs, the two males face each other. They then clash, with their heads lowered.
Above right: The distribution areas of the buffalo at the beginning of European settlement (shaded area) and today (red areas).

Opposite page: Two prairie dogs are caught by the photographer in front of the entrance to their burrow. Prairie dogs live in large groups called "colonies" or "towns." One group may include more than five hundred prairie dogs.

some 3 feet (1 m) tall, weighs up to 130 pounds (59 kg), and lives only in North America.

Both sexes have horns which have a bony core and a horny (protein) sheath made of hard keratin. Keratin is the same substance of which hair and nails are made. The sheaths are shed each year, in October.

The fur of the pronghorn can be light brown to dark gray. Its hairs are hollow, filled with air. This structure provides excellent protection against the cold winters. Also, these hairs can be raised, and during the summer they are raised to help cool the body. But during the winter they are held tight against the body to help insulate.

The pronghorn's eyes are larger than a horse's and bulge from the sides of the animal's head. This gives the pronghorn excellent eyesight and a wide angle of vision. Very few predators can get close to a group of pronghorns without being seen, and even if they get close enough, the pronghorns can run long distances at speeds over 37 miles (60 km) per hour. As they flee, pronghorns raise a patch of white hairs on their rump, thus giving the other members of the herd an easily seen warning signal.

Pronghorns can be seen at any time of the day. They

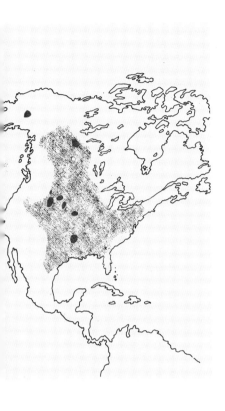

move in small groups, grazing on sagebrush, grass, and other small plants. The young are born in May or June and, at first, stay hidden in the bushes. There, from time to time, they are fed and tended to by their mothers.

Prairie Dogs, Squirrels, and Hares

A typical dweller of the plateaus, usually found in the Rocky Mountains or more to the east, is the black-tailed prairie dog. It belongs to the squirrel family, and lives in large communities or "towns." Stories are told about a town, at the turn of the century, which was 19,300 square miles (50,000 sq. km) large and contained 400 million animals. Obviously, there is no longer any way to be certain, but maybe, before agriculture upset wide areas of the plateau, some prairie dog communities could have approached this size.

A prairie dog town is a maze of tunnels. Each prairie dog defends a territory within which a system of burrows stretches for 9 to 90 feet (3 to 30 m) and to a depth of about 12 feet (4 m). The tunnels connect two or three openings, each surrounded by a small mound of dirt. One of these mounds

Except for the months of April and May, when the females are pregnant and give birth to their young, the openings of prairie dog burrows can be used by all members of the community. The animal wishing to enter need only be identified by another individual who might be at the entrance. The complex identification ritual, shown here, is ended with a "kiss."

is only one foot (30 cm) tall, but about 6 feet (2 m) wide, while the other has steep walls about 3 feet (1 m) tall and is shaped like a crater. Underground, along the tunnels are chambers, or rooms, used to raise the young or to store food. Other branches of the system of burrows can be closed, or may be used for burying droppings or dead animals.

Usually, all work around the openings is done soon after it rains, when the soil is moist and can be easily heaped to form mounds. These mounds, built around the burrow openings, are thought to be observation points, from which predators can be seen while they are still far away. Also, they reduce the risk of floods in the tunnel system during violent summer storms.

Inside the tunnels, the climate is more even than on the ground surface. In summer, for example, it is cooler and more moist. This is helpful, since in the summer it can become very hot in the grassland. The two entrances, one with a high mound and the other with a low mound, provide a continuous air flow, created by convection currents through the tunnel system. The high mound acts like a chimney, causing an updraft of air. Continuous air circulation through the burrows is important to avoid the accumulation of carbon dioxide during long periods when the animals are underground. In winter, the tunnels are a shelter against a snowstorm that may rage outside. The prairie dogs stay active throughout the cold season, and use up the food supplies of grass and other plants which they have collected during the summer. Besides the food supply, though, their body fat is also important, especially when spring is late in coming, and little food is left in the chambers.

The social system of the prairie dogs is very complex. Every town is divided into "neighborhoods" or "districts." These are marked by natural boundaries like rock piles, stream beds, and so on. Inside each district, there are "circles," each of which is headed by a male and contains many other males, females, and young. The members of a circle greet each other with a "kiss" that probably is a way of individual identification. All intruders, on the other hand, are immediately expelled, sometimes quite violently.

Prairie dogs feed on grass and, in summer, on other plants, rounding out their diet with grasshoppers and other insects. While the members of a colony are eating, one individual acts as a guard on top of a mound. In case of danger, it will give out a kind of two-syllable bark, which can be repeated up to forty times a minute.

The jackrabbit is not actually a true rabbit. It is a hare. Like the one pictured here, hares have slender bodies and long ears and legs. The jackrabbit, in fact, is the most extreme of all hares, and has all of the hare-like features at their best.

When threatened by a predator, a skunk will lift its tail as a warning. It will sometimes even stand on its front legs. If necessary, the skunk will spray a bad-smelling liquid. Besides the smell, if the spray hits the enemy in the eyes, it can even cause temporary blindness.

The pups are born in March or April, and stay inside their dens for about six weeks before coming out into the open for the first time.

Prairie dogs are not common west of the Rocky Mountains. Here there are several species of ground squirrels, which live in the grasslands and in the lower parts of the valleys. Typical examples are the Townsend ground squirrel and the Washington ground squirrel. Both have reddish gray fur, but the Washington ground squirrel's fur is also speckled with light gray.

The ground squirrels do not live in large towns like prairie dogs, but form scattered colonies instead. When winter comes, they hibernate. If the climate is not too extreme, though, they will emerge from their burrows in January or February. The ground squirrels, like prairie dogs, feed on grasses and other plants. During the hottest summer days, when the temperature on the ground can reach 122°F (50°C), these animals go into a second hibernation This period is called "estivation," or summer dormancy. Estivation can last up to three months (May, June, and July) in the hottest areas, while in more northern regions it may last only a month or a little more.

Another frequent dweller of the grasslands and plateaus is the white-tailed jackrabbit. Actually, this animal is not a rabbit, but a hare. The young of hares are born with fur and open eyes, and they can run. The young of rabbits are born without fur, with closed eyes, and they are helpless. The jackrabbit feeds on grass and other plants during the summer, and on twigs and bark in winter. It is mainly nocturnal, and spends the day sitting among rocks, in a bush, or in the grass. It does not hibernate, and during the winter it can dig tunnels in the snow to reach the grass.

Jackrabbits are amazing runners. They have been timed running at over 37 miles (60 km) per hour. The young are born in spring and summer.

The Skunks

Stretches of plateaus, grasslands, and sagebrush are crossed by streams and rivers. Areas which are always supplied with water, especially in the most sheltered valleys, have thicker and more lush vegetation. Here are found trees like the cottonwoods, so called because of their hairy and fluffy seed capsules. These "oases" of lush vegetation have a varied community of plants and animals. One member of this community is the skunk, a carnivore which can be found all over the region. It has a distinctive appearance, due to its thick black-and-white striped fur, and to its slow and awkward walk.

The skunk is slightly larger than a cat, and, although mainly nocturnal, it is easily found by its smell. Actually, its showy black-and-white striped fur is a warning signal to possible predators. If one of them dares come too close, the skunk gets ready to counterattack in a very odd way, by turning its rear side toward the enemy. If the predator persists, the skunk sprays it with a secretion from special glands, located near the base of its tail. This liquid has a terrible smell. In large doses, it can completely repel any predator, causing shock and vomiting. Also, the smell is very long-lasting. A sprayed animal will smell like a skunk for several days or even weeks.

Usually, this defense system is highly effective. But some predators, like the great horned owl, seem to be immune to it, and successfully prey on skunks. Some scientists think that this owl's sense of smell is so weak that it does not smell the skunk's strong odor. It is also possible, though, that the large flying predator attacks the skunk suddenly and by surprise, and kills it before it can spray its smelly liquid.

Skunks are omnivorous animals. They feed on small rodents, birds and their eggs, insects, larvae, berries, leaves, and sometimes also carrion. They usually live alone, but sometimes several females live in the same burrow, especially during the winter. At this time, in the northern part of their distribution area, skunks come out of their burrows occasionally to look for food. In the southern regions, on the other hand, they stay active throughout the year.

The young are born in the spring, naked and with their eyes closed. They stay in their den for about two or three months. Then they will start to follow their mother during her night trips, typically moving in single file, with the parent in front.

Predators

Large numbers of hoofed mammals and smaller animals, like squirrels and jackrabbits, live in the plateaus and grasslands at lower altitudes. They are a good food supply for many predators, which, in turn, keeps the prey populations in balance. Sometimes even mountain lions reach the plateaus, but usually they stay in the hard-to-reach canyons. Over a century ago, packs of wolves and lonely grizzly bears

The great horned owl is one of the largest nocturnal birds of prey in North America, and in the entire Northern Hemisphere. Other than skunks, it preys on opossum, porcupines, domestic cats, ducks, and other birds. This bird lives all over the continent and in some areas of the extreme north.

roamed the plateaus. Today, these predators only survive in the northernmost part of the region and, in very small numbers, in the most remote parts of the national parks. Some smaller species, though, are still very common today.

The American badger, for example, is a typical dweller of plateaus and lower areas covered by sagebrush. It is a sandy gray color, and it has black-and-white stripes on its face. It can be 2 feet (50 cm) long, and weigh 24 pounds (11 kg). Despite its rather small size, it is extremely strong, and its paws are armed with powerful claws, well fit for digging. The badger is truly a very skilled digger. While it digs its hole, it looks almost as if it is "swimming" into the ground. A strictly nocturnal animal, it preys mainly on ground squirrels, prairie dogs, and other rodents, which it reaches by digging up their burrows.

At the edge of the plateaus, where the forests begin, the red fox is found. Small and nimble, it looks more like a cat than a little dog. Actually, this species resembles a cat in many ways. It has small, sharp canine teeth and hairy footsoles. The pupils in its eyes are vertical slits. The fox's hunting behavior is also somewhat similar to a cat's. It performs acrobatic jumps into the air, or even up onto the

The American badger belongs to a different genus than the European badger. The two species, though, are similar in some general features, like the stocky body, short legs, thick fur, and tough skin. The skin is "oversized" in relation to the body. Badgers are incredibly strong and excellent diggers.

83

The whistling swan is the most common wild swan in North America. It is often seen along the Pacific and Atlantic coasts, in wildlife reserves and on inland lakes. It is similar to the trumpeter swan, but, is smaller, with a yellow spot at the base of its beak. Its call is similar to a goose's.

lower branches of trees. It is a very skilled hunter of rabbits, ground squirrels, and other rodents, but it also eats berries and other fruit when it can find them. The fox is mainly nocturnal, but it can also hunt during the day, especially when it has to catch food for its young. The mother cares for the young alone, while the male catches prey for both. Red foxes usually have two or more dens, so that the pups can be moved quickly if their refuge is disturbed.

The most common predator in the region, and probably also the most successful, is the coyote. It is a medium-sized animal, 3 feet (1 m) long and weighing 40 pounds (18 kg). This species has almost completely filled the vacant niche following the almost total extinction of its close relative, the wolf. Today, it is common all over the region, especially in the lower basins and on the plateaus. It prefers hunting at night, but can be active any time of the day. Coyotes feed on all types of food, including berries, other

Following page: A group of white pelicans rests on old tires in a lake in Klamath County, Oregon. These large birds migrate in huge "V" formations, and often soar with the wind at high altitudes. Unlike brown pelicans, these birds do not dive under the water to catch their fish.

fruit, rodents, birds, and carrion. They do not hunt in packs like wolves, and so cannot kill large prey. Still, they will sometimes hunt old or sick deer, and especially their fawns.

Often, in grasslands, coyotes can be heard at sunset as they send out a sequence of barks and howls. These are not as deep and long as a wolf's, and they are interrupted by haunting laughing sounds. The coyote's howl is among the most typical sounds in the region.

Swans and Pelicans

In spring, large numbers of birds fly across sagebrush basins and grassy plateaus during their migration to their nesting places, the northern forests, and arctic regions. A great many of them also stop to nest in the region, and over three hundred species are frequently observed. The largest variety of bird species is found along the rivers and near lakes. One of the most spectacular species is the trumpeter swan. In the past, this species nearly became extinct, but now its population is growing again. Small, isolated groups of trumpeter swans nest along the rivers and by the lakes and marshes of Yellowstone National Park. They are also found in some hard-to-reach regions of Montana and Wyoming, and in the Canadian Rocky Mountains.

The reproductive season for this large swan is very short on the plateaus, where the ground is sometimes free from snow and ice for only eighty days a year. At the beginning of the season, sudden temperature drops can cool off the eggs, and even kill the growing embryos in them. Autumn frosts often come very early in August, and can seriously injure the young. Frost kills water plants, which supply the swan with food, and allows predators to draw close, since the rapidly formed ice crust is easy to walk on.

During the winter, most of the swans move to warmer regions, like Puget Sound. A few stay on the cold plateaus, especially in Yellowstone National Park. Here, hot springs keep some streams and marshes from freezing. In winter, these swans eat roots of water plants from the bottom of the lakes, and sometimes also fish and other small animals. Their long neck allows them to reach down to the bottom, in water 3 feet (1 m) deep.

Usually these birds live in pairs or family groups, but in winter they may gather in groups of up to thirty or forty. During the mating season, the pairs become very aggressive. The birds defend a territory of 5 to 25 acres (2 to 10 hectares), depending on the amount of food available. In spring, the

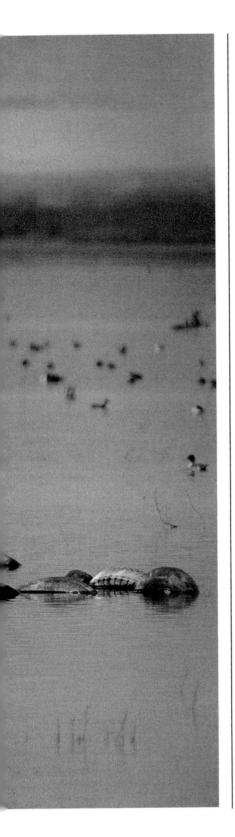

loud call of the swans trumpets from the water. It is a continuous territorial warning to other swans. The call, resembling a trumpet sound, originates from air being pushed through the trachea, or windpipe. The trachea is long, winding, and specially shaped to produce sounds. The trachea touches some chest bones, which vibrate, and help create the deep trumpet call.

Another bird, found in the most isolated lakes of the plateaus, is the white pelican. It is not very rare, but it is a dweller of the wildest areas. It nests on larger marshes (for example, on Yellowstone Lake). Like the trumpeter swan, the white pelican is a huge bird whose wingspan can even reach 9 feet (3 m). Its plumage is almost all white, with two black spots on its wings.

The large pelican's bill is especially adapted for fishing. It is over 8 inches (20 cm) long, and is equipped with a large throat-pouch of skin. When the pelican fishes, it will gather a lot of water in its throat-pouch, and some small fish with it. Then it releases the water, pointing its bill downward and squeezing the pouch against its chest. The fish remain inside the pouch and are then swallowed.

White pelicans are interesting to scientists, because of their group fishing habits. Several birds, up to ten or more, fly in a line over shallow waters, "pushing" the fish forward and then closing into a circle to catch them quickly. This "group work" is very effective, and can be repeated for hours.

The large bill and the hanging throat-pouch also have other functions. The skin walls have many blood vessels, which expand in summer. This allows excess body heat to escape through the wide pouch surface. The heat release is helped by the so-called throat beat. The throat-pouch moves in and out at a rate of up to 250 beats per minute. Measurements of body temperature on captured pelicans have shown that the "throat beat" can lower the temperature by 41°F (5°C) in a few minutes.

White pelicans nest in colonies, building large nests with twigs on the ground. At the beginning of the mating season, they grow disk-like knobs on the upper part of their bills. This strange structure seems to be meant to attract a partner during the courtship. It can also stimulate a rival's anger, and is often a favorite target during the frequent territorial fights which occur among pairs inside the crowded colonies. During such fights, over 95 percent of the pecking is aimed at the knob. Thus, it reduces the risk of serious wounds from the powerful bird's bill.

Top: A sandhill crane wanders in a meadow in the Grand Teton Range, Wyoming. During their migration, flocks of these birds drop from the sky and stop in fields, open grasslands, and in marshes to feed on insects, small rodents, frogs, fish, and small water invertebrates. Despite their large size, their dull plumage blends well with the background, and they are hard to see. When near, though, the large red spot on the bird's head is quite plain.

Right: This drawing shows both adult and young whooping cranes. This is one of the rarest species of birds in the world. Less than a hundred individuals still live in the wild.

Opposite page: The map shows the areas where the whooping cranes nest and spend the winter—present and past—and the direction of their migrations.

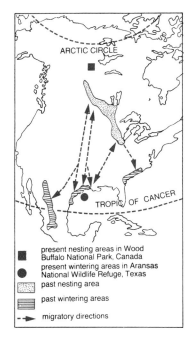

present nesting areas in Wood Buffalo National Park, Canada
present wintering areas in Aransas National Wildlife Refuge, Texas
past nesting area
past wintering areas
migratory directions

During the nesting period, when territorial fights are over, the knob shrinks in size and eventually disappears. Normally, two eggs are laid, and they are brooded in quite an unusual way. Some birds develop "brooding plates" on their chests and abdomens. These plates, which contain many blood vessels, supply much heat to the egg. The pelican, however, does not develop a brooding plate. Instead each egg is covered by a large, webbed foot. The feet, like the brooding plates, have many blood vessels, so they are a good source of heat.

Often both eggs will hatch, but only one chick reaches the age of flight. The first hatched gets much more food, so it grows much faster. The unlucky second born is doomed to starve to death, or to be pushed out of the nest by its older brother or sister. Once out of the nest, the little bird is no longer recognized as one of the family, and it is soon attacked and killed by the adult birds, often by its own parents!

Common Cranes and Endangered Cranes

In spring, migratory flocks of sandhill cranes cross the moist areas. They come from their wintering regions, in the southern part of the United States. This is another large species, with a wingspan of some 6 feet (2 m). Its neck and legs are extremely long, so that this bird is over 3 feet (1 m) tall. Most of the birds do not stop, and head on to the regions farther north. Some, though, stay and nest in the most secluded swamps, grasslands, and pools in the open forests. Often it is possible to hear the sound of the cranes in flight, even before seeing them. This is a pleasant musical interlude, one of the most awaited and welcome spring voices.

Among the many special features common to cranes are their mating dances. The two partners face each other, give out a typical call and jump up into the air several times, spreading their wings open. At the climax of this enraptured ritual, they collect grass and throw it in the air, jumping one more time. Later on in the season, the couple becomes much more secluded, and builds a bulky nest in the thick of the lake vegetation. In it, one or two eggs will be laid.

The young cranes are covered by a brownish plumage and look quite different from the adults. They keep their coloration for a year, to the delight of biologists who are trying to evaluate the reproductive success of the cranes. All they have to do is count the young born during the year, when the flocks have reached their wintering areas.

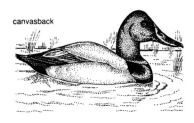
canvasback

western grebe

yellow-headed blackbird

killdeer

Another large bird in the wet environments is the whooping crane, which is even larger and more spectacular than the sandhill crane. This bird is mostly white, with partially black wings, a black mask around the eyes, and a bright red top of the head. Once this crane was rather common throughout the region, but hunting and the destruction of its habitats have taken it nearly to extinction. Today it nests only in Wood Buffalo National Park, Alberta, Canada. The entire population spends the winter in or around the Arkansas National Wildlife Refuge, on the coast of Texas. A few years ago, there were only twenty whooping cranes left. Today there are more, but the species is still seriously endangered.

In their efforts to save this beautiful bird, biologists have even tried to put fertile whooping crane eggs into the nests of sandhill cranes, which are much more common and closely related. This experiment has been going on for many years with the help of a population of sandhill cranes nesting in Idaho. The adoptive parents have successfully raised many whooping cranes, but scientists cannot yet be sure that these birds will reproduce, starting a new group of whooping cranes in Idaho. The main problem is that the early experience of whooping cranes with their adoptive parents (called imprinting) might lead them to prefer a sandhill crane to a whooping crane when mating. The result might thus be a population of hybrid cranes, instead of a new population of whooping cranes.

Other Birds in Wet Environments

Rivers, lakes, and other wet environments also have many other bird species. Among the ducks, for example, there are the mallard, blue-winged teal, cinnamon teal, and shoveler. The shoveler stirs up the bottom silt with its large, flat bill, looking for small invertebrates.

The diving ducks such as redheads, canvasbacks, ring-necked ducks, and the ruddy ducks are common.

Among the grebes, one of the most common species is the western grebe. Like the other species in its family, this grebe has a very elaborate mating display, with offerings of water plants, dancing on the water, and so on. Its floating nest is made with lake plants, and is attached to reeds or to a branch stuck in the bottom silt, always close to the shore or in shallow waters. Songbirds are also common in this environment. The red-winged blackbird and the yellow-headed blackbird nest in the reeds in colonies. The male red-

The American avocet is one of the most peculiar water birds of North America. During the nesting period, its neck and head are an attractive cinnamon (reddish brown) color.

winged blackbird engages in much territorial behavior, and continuously warns rivals by showing the bright red patch on the upper part of its wings and giving out a loud scratching sound. The males of the second species, on the other hand, mark their territory with a peculiar song, which is somewhat like static on a radio broadcast.

Males of both species try to establish a harem of females. Rarely will a male yellow-headed blackbird have more than two females, while males of the red-winged blackbirds can have more than ten females.

In the marshes there are also several species of herons, rails, and other marsh birds. The great blue heron is quite common, but visitors may also come across the American bittern, the green heron, and, in the southernmost parts of the region, also the small least bittern. It is always very hard

The shoveler is widespread, not only in North America, but also in Europe and Asia. The American populations of shovelers nest in a wide area in northwestern North America, and winter in Mexico and in the southern United States from California to Florida. In the picture, a female shoveler is seen.

Among the most typical birds of prey of the American Northwest are the rough-legged hawk, osprey, marsh hawk, and goshawk (found also in Europe, Asia, and sometimes on other continents), and birds native to North America, like the red-tailed hawk, sparrow hawk, turkey vulture, and the ferruginous hawk.

to spot the rails, since they stay constantly hidden in reeds, rushes, and thick vegetation. In the spring though, the call of the males is often heard, and the deep grunt of the Virginia rail is one of the most familiar sounds in the marsh.

Among the marsh birds, the American avocet is a very typical species. It has black-and-white plumage, a chestnut head, and a long bill, curved upward. It lives in most of the muddy swamps and pools throughout the region. The killdeer is also quite common around pools, especially on dried-up silt and pebbly areas. This bird is a typical plover, with a brown back and two black stripes on its chest. This species has been named after its call, a loud and repeated kil-deer-kil-deer.

In the open fields and grassy plains which surround the

marshes, the long-billed curlew can be found. In spring, the males, which control a territory, fly in wide arches, singing a pleasant song. Finally, in some parts of the region (Washington, Idaho, and Montana), the upland plover makes its home. Unlike most of the marsh birds, these birds nest on plateaus and high pastures.

Birds of Prey

In the region between the mountains, over twenty species of birds of prey have been regularly recorded. Several species are quite common, and feed on rodents, small birds, and insects. The smallest daytime bird of prey, the sparrow hawk, can often be seen along roads, perched on telephone wires. It carefully scans the ground below in search of

A male sparrow hawk (in front) with two young, at the entrance to their nest. A typical representative of the falcon family, this bird of prey is only 8 inches (20 cm) long. It weighs less than 4 ounces (113 g) and is the smallest bird of prey of the Northern Hemisphere. It is often seen perched on telephone wires or on trees along the roads. It carefully surveys the ground and branches below, looking for small rodents, insects, or small birds. Usually, though, it hunts in flight, with frequent hovering. Despite its small size, it is a genuine bird of prey, capable of hunting prey almost as large as itself. It nests in tree holes, in cracks in the walls of buildings, or in crevices in rocks.

rodents or insects, like beetles and grasshoppers. This hawk often also catches sparrows or other small birds. For this reason little songbirds are quite afraid of it. They will try to attack it and chase it away if one gets too close to their nests. Robins and mockingbirds are often leading the violent and loud attacks against the sparrow hawk.

The main daytime bird of prey is the golden eagle. This bird is never common, but can still be found in open grasslands, along canyons, and in the mountains, above the timberline. This eagle mainly preys upon large rodents like marmots, ground squirrels, and so forth. In some areas of the Rocky Mountains, golden eagles have recently been subject to a peculiar and shameful kind of poaching. The birds are chased by airplanes, and shot down with heavy gunfire from a short distance, in full flight.

The buteos or soaring hawks are more common than the eagles. Five species are found in the western regions of North America. All have large bodies, relatively small beaks and claws, and rather rounded wings and tails. One of the most typical species of the West is the ferruginous hawk. Its plumage can be either very light, almost white, or dark. The lower part of the body is an even brown, except for the wing and tail feathers, which remain very light.

At night, nocturnal birds of prey take the place of the daytime predators. Among them, the smallest is the flammulated owl, a migratory species. This bird prefers the open ponderosa pine forests, bordering grasslands or along ridges. It is a shy bird. It is also hard to locate, because its call is soft and seems to come from another direction.

The main nocturnal bird of prey in the region is the great horned owl. This owl has a wingspan of 4 feet (1.4 m) and powerful claws. It hunts large rodents, and also more aggressive animals, like skunks and, occasionally, smaller owl species. It is easy to recognize by its large size, but also by the feather "horns" which rise on its head. During the day, its large, yellow eyes are easily visible. This bird lives in several habitats, like canyons, open grasslands, forests, hills, and mountains.

Grouse of the Sagebrush

Another common dweller of sagebrush areas is the large and shy sage grouse. This is one of the better-known American species belonging to the Tetraonidae family. It feeds mainly on leaves and twigs.

In spring, the males gather to display on a "lek," which

female sage grouse

The drawings show some steps in the courtship ritual of a male sage grouse. From the beginning resting position (top, left), the male begins to lure the females, who are wandering in the lek. The male puffs out its gorget ("collar") of white feathers around its chest. It also fans out its pointed tail feathers, partly lowers its wings, moves sideways with short, rhythmic steps, and inflates its chest sacs (top, right, and bottom, left). At the climax of the ritual, the male's wings are loosely hanging down, its head has almost disappeared among the feathers on its gorget, and the air sacs are enormously inflated (bottom, right).

is a flat area on grassy, open ground. Usually, the same leks are used year after year. Some of the largest leks, located in the most hard-to-reach areas, have probably been used for many years. On the lek, each male defends a small area, and every morning tries to prove his dominance over the other males in order to win the females' favor. The females draw near to mate with the males which have the best areas on the lek. Some leks will hold only four or five males, but some others may be used by fifty birds or more.

During the display, the male inflates two skin sacs located on its chest with air and lifts an area of white feathers around its chest and neck. It fans out its tail, showing its beautiful pointed feathers. These form a kind of background for the body. The bird shakes up and down, giving off a deep, booming call.

The females wander among the displaying males, seeming to pay no attention. Actually, they are rating each male's quality, and choose to mate only with certain males. On some leks, one or two males alone may engage in up to 90 percent of the matings. Scientists still do not know what features the females prefer to select. Certainly the social position on the lek, and the quality and frequency of displaying are of some importance. After mating, the females move out of the lek and go to lay their eggs, brood them, and raise the young without help from the males.

Another rather common bird, related to the sage grouse, is the sharp-tailed grouse. This species is smaller than the sage grouse and prefers habitats with open meadows. The males of this species also display on a lek and, although their plumage is not as spectacular as the sage hen's, their behavior is even more interesting.

On the leks, the males hop around in a circle and briefly chase other males. Their peculiar walk resembles the movement of a wind-up toy. As they move about, they keep their tails up, showing their white lower feathers, and inflate two pink skin sacs at the sides of their necks. The display is accompanied by a deep cooing, like a pigeon's call. The females, as with the sage grouse, visit the leks to choose their mates. They then go back to the shrubs to brood their eggs and raise their young alone.

Reptiles and Amphibians

The most widespread and best known reptile of the region is the western rattlesnake, which is rarely over 3 feet (1 m) long. The rattlesnakes have a series of scaly sections at

A western rattlesnake suns itself in the midday sun. The rattlesnakes have large poison fangs in the front of their mouths. These fangs are hollow and can be folded backward. Their poison, contains a substance which destroys blood vessel walls, another which causes blood coagulation, and others which destroy white blood cells and cells of other tissues.

the end of their tails. These sections interlock, but stay rather loose and, when shaken, produce the well-known rattling sound. This sound is a warning signal to any possible predators, or animals, like large herbivores, which could step on the snake. The young rattlers may add three or four sections a year to their rattles, usually when they shed their skins. Older rattlesnakes add one or no sections to their rattles each year.

All rattlesnakes are poisonous and very dangerous. They have a very effective way to inject their poison. In the front part of their mouths, they have two hollow fangs which

The spadefoot resembles a small common toad. The spadefoot, however, has a shorter head and eyes with vertical, instead of horizontal, pupil slits. In times of drought, this tiny animal spends weeks or even months hidden in the cool soil. It can swiftly bury itself with movements of its hind legs, thanks to the "spades" it has there. Its 1,000 to 1,500 eggs, laid in ribbons which are wrapped around water plants, hatch in only two to fifteen days.

can be folded inward and are raised when the snake bites. These are forced into the victim's body. The poison is then pumped out of the poison glands with special muscles and flows through the hollow fangs. It has a very rapid effect and kills small prey, such as rodents or birds, within a few minutes.

Rattlesnakes also bite to defend themselves, so they can be very dangerous to people. Western rattlesnakes are usually rather small, but their bite can kill an adult in good health, and it is extremely painful. The victim who does not die, will be in severe pain for several days, unless he or she is immediately given an antitoxin, which neutralizes the poisonous substance.

In the middle of summer, when temperatures get higher than 104°F (40°C), grasses turn yellow and brown, and the plateaus become dry and dusty. It is hard to believe that, in such conditions, various species of amphibians can survive, and even thrive. Actually, frogs, toads, and salamanders can all survive if, of course, they keep close to marshes, pools, and rivers.

One species of toad, the Great Basin spadefoot, is at times very common, and it can reproduce even in the most short-lived pools. This tiny animal, usually no longer than 1 or 2 inches (2.5 to 5 cm), has vertical pupil slits, and is equipped with a hard, sharp "spade" on each hind leg. It uses these spades to dig in the soil, with rotating movements of its hind legs and its whole body. This way it gradually buries itself in the ground. It may spend several days in its burrow, coming out at night or in damp weather.

The spadefoot has rather smooth skin and small teeth on its upper jaw. This is another peculiar feature of the spadefoot. These teeth can release a poison when the toad is handled and for this reason, most predators avoid it.

These small amphibians have "explosive" reproductive habits. In most of the areas where they live, spring and summer temperatures and rainfall may be up or down. When enough rain falls on the plains, these creatures can literally fill up pools and puddles, and within a few hours their "choir" (mating calls) will be heard several miles away. Soon they lay their eggs, attaching them to water plants. The tadpoles grow very rapidly and are able to leave the pools within a few weeks, before a boiling sun dries them out.

GUIDE TO AREAS OF NATURAL INTEREST

Opposite: A lone fisherman views a mountain area in northwestern North America. Spectacular mountains, large conifer forests, rivers, and lakes are scattered all over in the American Northwest. The traveler seeking wilderness will find some of the most interesting environments in the world here.

Following page: The areas of natural interest in the American Northwest.

Below: The area dealt with in this volume: Canada and the United States.

The Rocky Mountains and the Pacific Northwest region are rich in parks, national monuments, recreation areas, national forests, wilderness areas, and national wildlife refuges. All of these various kinds of protected areas are open to the public, even though a four-wheel drive vehicle is often necessary to get to the most hard-to-reach areas, or long hikes may be required. Most of the areas are easily reached by paved roads, and in many cases there are camping facilities. Usually there are motels and restaurants in nearby towns.

A complete list of all the areas of natural interest would fill an entire book. Thus, only some examples which are typical of the region's main habitats will be listed in this section. Visitors will find most of the plants and animals mentioned within this book in the areas listed below.

Some of these are truly wild areas, in many cases never modified or even touched by humans. It is thus important that the visitor be prepared to deal with bad weather and with other possible inconveniences. Whenever planning a long hike, it is absolutely necessary to bring along a compass and a good topographical map and to know how to use them. It is also a good idea to ask the local park official for information pertaining to trails and the possible risks to the trip. They should have the destination of your hike in writing, the plan for getting there, and the estimated return time. Along most of the hardest trails, there are "registration stations," where it is customary to sign in before and after each trip. In some parks, this procedure is required.

When planning to sleep out, it is necessary to bring along enough food and water, a flashlight, and a first-aid kit. Also, you should remember that in flat and arid areas, summer temperatures can rise over 104°F (40°C), and there might be no water within several miles. On the mountains, both in winter and summer, sudden and unexpected storms might hit the area, with sharp temperature drops.

Also, before zipping up your sleeping bag for the night, be sure to hang up all your food with a rope, at least 13 feet (4 m) off the ground. Bears have often robbed careless campers overnight, not to speak of the frightening experience of having an adult bear trying to get into the tent. All this does not mean, though, that the parks and protected areas listed below are dangerous and hard-to-reach places. On the contrary, most of the visitors have had unforgettable experiences and, with a little organization and common sense, accidents can be avoided.

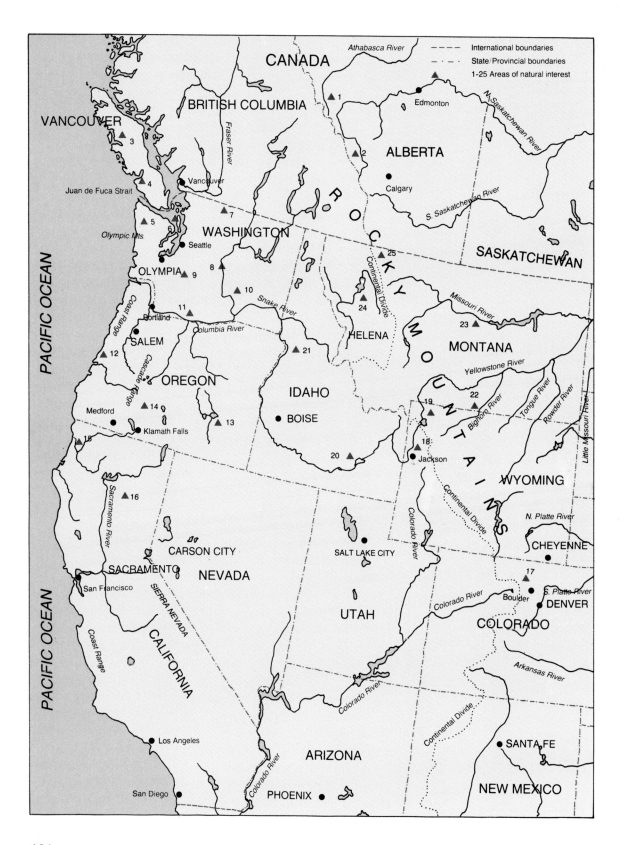

CANADA

Alberta: National Parks Jasper (1) and Banff (2)

The western regions of Alberta province are dominated by the northern part of the Rocky Mountains. There are many national and provincial parks, totaling about 5,018 sq. miles (13,000 sq. km) of land. In some cases, these parks extend across the Alberta-British Columbia border.

Two famous and spectacular parks are Jasper and Banff. They are both easily reached by paved roads, from Edmonton or Calgary. The mountain landscapes are truly breathtaking, with huge rock walls, glaciers, snowfields, and alpine meadows. Blossom time is June through August, depending on the altitude.

The fauna is often very interesting. In the forest, besides mule deer, there are wapitis and caribou. Among the carnivores, there are mountain lions and bobcats, but these animals are seen very rarely. At timberline, marmots, pikas, bighorn sheep, mountain goats, grizzly bears, and wolverines can be found. Among the birds, on the alpine grasslands, the white-tailed ptarmigan and rosy finches are common, and eagles and peregrine falcons are often seen.

British Columbia: Stratchona (3)

This provincial park, placed in the middle of Vancouver Island, covers about 463 sq. miles (1,200 sq. km). It includes areas of forest and mountain peaks which form the backbone of the huge island.

At timberline, one can find a native species, the Vancouver marmot. During the summer, alpine flowers are common, and various birds can be seen. Among them are the golden eagle, the bald eagle, the pine grosbeak, the evening grosbeak, and also red crossbills, olive-sided flycatchers, white-tailed ptarmigans, common ravens, gray jays, and varied thrushes. With a bit of luck, a pileated woodpecker can be spotted. Actually, the whole of Vancouver Island is one of the best spots for bird-watching in North America. Over two hundred species of birds have been recorded there. Other than the forest species, there are grebes, loons, seabirds (common along the coast), ducks, geese, swans, herons, and other marsh birds. Two unusual birds are the meadowlark (which has been introduced from Europe) and the Chinese myna, from China. Among the mammals, there are mountain beavers, Douglas' squirrels, mule deer, black bears, chipmunks, and martens.

British Columbia: Pacific Rim (4)

A stretch of coast about 43 miles (70 km) long, on the southwestern corner of Vancouver Island, has been made into a national park. It is possible to drive to the park from

Watson Lake, British Columbia shows off its autumn colors. The shows offered by nature in the protected areas of the American Northwest and Canada are still incredibly beautiful today.

Victoria and Nanaimo, taking the ferry from Vancouver, and then Highway 4 through Port Alberni.

Close to the high-tide line, the rain forest begins, with sitka spruces beautifully covered with mosses. Many seabirds can be seen along the shores and on the rocky coasts. Among them, there are cormorants, birds of the Alcidae family, sea ducks, gulls, and, during migration season, various marsh birds. The river otter can also be found along the shores, especially near the mouths of small streams. In spring and in autumn, water jets several feet high may sometimes be seen out toward the open sea. These are caused by gray whales coming to the surface to breathe.

UNITED STATES

**Washington:
Olympic National Park and coastline (5)**

This national park includes most of the Olympic Mountains and the rain forest along the coast. Some stretches of forest along the major rivers are also protected. Within the park there are about 1,930 sq. miles (5,000 sq. km) of forests, with giant conifers, snowfields, glaciers, and barren ridges.

The Olympic Mountains are reached by car from Seattle or Portland, but the higher peaks can be reached only on foot. There are several access points to the coast, all paved roads (for example, from Neah Bay, La Push, and Queets). Starting from there, the most remote beaches can be explored only on foot. Excellent trails wind up the mountains. Thick forests cover the valleys.

The fauna includes black bears, wapitis, mule deer, mountain goats, and the native Olympic marmot and pikas. The mountain lion is common in the area, but very difficult to see. Among the birds, blue grouse, crossbills, hawks, and ravens are common.

Interesting trips can be organized along the beaches. Stretches of sand alternate with rocky banks and tide pools. During the migration period, marsh birds are sometimes very common. Offshore, on the open sea, shearwaters and whales can be seen. A colony of tufted puffins sometimes nests on a rock wall by La Push harbor, and often there are mergansers along the river. Bears, river otters, and sometimes also bald eagles live in the coastal forest. More inland, beautiful trails lead to the middle of the rain forest.

**Washington:
Puget Sound (6)**

No large national parks exist in this area, but there are some small state parks and some national wildlife refuges which are fairly large and easily reached by car from Seattle or Bellingham. Skagitde Flat Game Refuge, for example, is one of the most interesting protected areas. It is a wide stretch of brackish marshes and wet fields. Here, many interesting birds can be seen any time of the year. In the middle of winter, large flocks of lesser snow geese live here, and from October through November there are falcons, red-tailed hawks, and owls.

Whidbey Island can be reached by Highway 20, starting from Interstate 5, north of Mount Vernon. Another way to reach the island is by ferry, from Mukilteo, near Everett. Especially interesting places on the island are Deception Pass and Rosario Beach. At high tide, the ocean covers the stretch of sand which connects Whidbey Island to Fildago Island, and rushes in at 6 to 12 miles per hour (10 to 20 km). The area is rich in bird life. Bald eagles, harlequin ducks, black oystercatchers, pigeon guillemots, marbled murrelets, and sea ducks live here.

The San Juan Islands also deserve a visit, but they are reached only by ferry, from Anacortes. The trip to the islands is very beautiful, past cliffs and small islands covered with forest. From the boat, sea ducks and cormorants can often be seen, and with a bit of luck, even common seals or some killer whales. Actually, a group of killer whales lives permanently in the area.

Washington: North Cascades National Park (7)

This national park, located in the northernmost part of the Cascades Range, south of the Canadian border, is one of the most secluded in the region. It covers about 965 sq. miles (2,500 sq. km), and includes huge mountains, glaciers, and snowfields. It is reached by Highway 20, which cuts across it, but it offers no tourist facilities. For this reason it is the ideal place for people looking for a real experience in true wilderness. The park ranger, though, can give good advice on the routes and possible risks. For example, even in summer there may be risk of avalanches. The plants and animals are typical of mountain forests and alpine meadows. Blossom time is in July and August. There are hoary marmots, pikas, black bears, mountain goats, wapitis, mountain lions, and wolverines. Among the birds, there are common ravens, gray jays, Clark's nutcrackers, white-crowned sparrows, and boreal chickadees. Along the upper course of the Skagit River, at the boundaries of the park, groups of bald eagles can be seen in winter. From November through April, up to a hundred eagles may gather here to feed on migrating salmon and trout.

This photo offers a view of Liberty Bell Peak, in the Cascades Range, near Washington Pass.

Washington: Wenatchee (8)

Most of the inland part of the state of Washington is heavily farmed, but some wildlife refuges and national forests have preserved large stretches of natural environments. On the eastern side of the Cascades, one of the places most deserving of a visit is Wenatchee. This town, on the Columbia River, is found in the border area between open forest and sagebrush grassland. Among the many interesting animals of the area, there are hummingbirds, western bluebirds, mountain bluebirds, yellow-bellied marmots, mule deer, and perhaps black bears.

Washington: Mount Rainier National Park (9)

This large national park covers over 232 sq. miles (600 sq. km) around the huge volcanic peak of Mount Rainier. It is accessible by paved roads, from Seattle or Portland, but its most remote parts are reached only on foot.

Large stretches of forests alternate with more open ground, with alpine meadows, bare rocks, and ice at higher elevations. This is the ideal place for observing a moving glacier at close range.

Black bears, wapitis, mule deer, mountain lions, chipmunks, Douglas' squirrels, and about seventy species of birds live in the forest. At the timberline golden eagles, white-tailed ptarmigans (especially at Panorama Point, at Paradise Lodge), rosy finches, Clark's nutcrackers, and gray jays are often seen. Among the mammals, hoary marmots, golden-mantled ground squirrels, and mountain goats are common.

Washington: Columbia and Potholes National Wildlife Refuges (10)

The Columbia and Potholes National Wildlife Refuges are reached by getting off Interstate 90, turning south just past Moses Lake.

In the area, there are numerous alkaline lakes, marshes of every size, dusty gulches called "coulees," and sagebrush plains. Among the fauna are coyotes, badgers, rattlesnakes, and various species of ducks. Yellow-headed blackbirds, red-winged blackbirds, sage grouse, and sharp-tailed grouse live in the drier areas. In spring, large flocks of migrating birds fly across the region, offering a beautiful show. In particular, sandhill cranes can be seen heading to their nesting locations in the Arctic.

Washington: Columbia Gorge (11)

This area can be reached by car on Highway 14 from Vancouver, or by Interstate 84 from Portland, along the Oregon side of the Columbia River.

Here the large Columbia River cuts across the Cascades.

Today, part of its beauty has disappeared due to the construction of a series of dams, but the forest and deep gorge are still splendid. Among the birds dwelling in the pine forest, there are wood warblers, thrushes, tanagers, and vireos. On the eastern side of the gorge, oak trees become more common.

Oregon:
coastline (12)

The over 621 miles (1,000 km) of Oregon coastline are certainly among the most breathtaking and well-preserved coastal environments on earth. Highway 101 runs almost the entire length of this area, connecting a long chain of small towns and local parks. The entire area offers motels and campgrounds where visitors can stop for the night. The coast is a sequence of beautiful views, and the fauna is very rich.

At Sea Lion Caves, to the south, visitors will see Steller's sea lions. The Oregon Dunes National Recreation Area is one of the few stretches of sandy beach in the area. In winter and spring, the numerous pools become filled with frogs and newts, which gather there to reproduce.

In spring and autumn, it is possible to see migrating gray whales off the coast. River otters live on the shore, birds of the Alcidae family are perched on the rocks, and opossums and raccoons live in the coastal forest.

Oregon:
Malheur National
Wildlife Refuge (13)

Malheur National Wildlife Refuge, in southeastern Oregon, covers about 2,702 sq. miles (7,000 sq. km) of land, with many lakes and marshes. In spring, natural history classes are held in the refuge.

In winter, many ducks and geese stop here, especially Canada geese and lesser snow geese. During migration time, sandhill cranes and white pelicans can be seen. To the north, in the mountain forests, live the pronghorns, but it is very difficult to get close to them.

Oregon:
Crater Lake
National Park (14)

This national park, which covers over 154 sq. miles (400 sq. km), includes the remains of a volcano which exploded thousands of years ago. Today, the crater is filled with water. Crater Lake lies at an altitude of about 6,562 feet (2,000 m) and can be reached by car from Medford or Klamath Falls. The slopes of the volcano are covered with conifer forests, and to the north, there are wide lava beds. The plants and animals are those typical of the southern part of the Cascades.

Among the woodpeckers, there are the white-headed

**California:
Redwood
National Park (15)**

**California:
Lassen Volcanic
National Park (16)**

**Colorado:
Rocky Mountain
National Park (17)**

**Wyoming:
Grand Teton
National Park (18)**

Opposite page: A curious raccoon peeks out of its den, in a tree hole.

woodpecker, Lewis' woodpecker, and the three-toed woodpecker. Among the other birds, hummingbirds, wood warblers, tanagers, and red-breasted nuthatches are typical. All of them are common in spring and summer. Among the mammals, there are black bears, wapiti, mule deer, golden-mantled ground squirrels, and martens.

This National Park covers about 96 sq. miles (250 sq. km), including some coastal areas in Northern California. It is easily reached by Highway 101, north of Eureka or south of Crescent City. Obviously, the principal features of the park are the huge and very old redwood trees. Guidebooks are available at the park center.

Lassen Peak marks the southern end of the Cascades. More to the south is the Sierra Nevada Range. The national park covers about 200 sq. miles (500 sq. km), and can be reached by car on Highway 44, from Redding.

The main feature of the park is its volcanic origin. There have been recent eruptions, and there are sulphur fumaroles and hot springs. Fumaroles are places where hot sulphur vapors rise from holes in the ground. Mount Lassen last erupted about eighty years ago, but ashes are still thrown out of the crater from time to time. Many trails wind around the mountain slopes, with good views of the various aspects of volcanic activity. The fauna is not very abundant, but there are some interesting species, like the Steller's jay and the golden-mantled ground squirrel.

This National Park covers about 232 sq. miles (600 sq. km). It includes magnificent mountains with high peaks, ridges, alpine meadows, and wide forests. The plants and animals are typical of the Rocky Mountains. The area is reached from Denver or Boulder, by Highway 36, or from Loveland by Highway 34.

The plants and animals of the huge mountain range of the Grand Tetons are not as rich and varied as in nearby Yellowstone National Park, but they include many typical mountain species.

The area is also popular for winter sports, especially around Jackson Hole. Besides downhill skiing, cross-country skiing is also possible, and in winter this is a good way to explore hard-to-reach areas, rich in fauna. Also, during the winter, large numbers of wapiti move down toward the

A group of curious tourists observes the slow road-crossing of a majestic male buffalo in Yellowstone National Park. After the mass slaughter carried out by the early settlers, this magnificent animal now lives in peace with people.

plains, northeast of Jackson (in the Elk National Wildlife Refuge). Buffalo live in the area as well.

Wyoming: Yellowstone National Park (19)

Located north of Grand Teton National Park, this is the largest of all North American National Parks, and the oldest park in the world. It includes about 3,088 sq. miles (8,000 sq. km) of mountains, sagebrush, grassland, lakes, and hot springs. Tourist facilities are numerous, excellent, and meet all needs. Visitors should remember, though, that only a few miles away from the roads, hotels, and campgrounds, the park is still completely wild, and hikers should take all possible precautions when planning hikes.

The geysers and the boiling mud pools which have made this park famous can be reached by a network of roads in the park. In the middle of summer, these roads are often packed with thousands of tourists. But the crowd thins out just a short distance from the paved roads, and one can soon be completely alone. Most of the park actually remains completely unknown to the average visitor. Also, in winter there are very few tourists, and the snow and superb ice formations created by geysers' jets freezing in the icy air provide a completely new show. This experience is unique, even though winters in the area may be incredibly cold. Often temperatures will drop to -40°F (-40°C).

Besides the breathtaking views and natural wonders, the park also contains very rich fauna. There are moose, wapiti, white-tailed deer, buffalo, grizzly bears, black bears, wolves, coyotes, bighorn sheep, pronghorns, marmots, golden eagles, ospreys, Clark's nutcrackers, common ravens, white pelicans, trumpeter swans, and many other species. All these attractions undoubtedly make Yellowstone one of the most notable areas of natural interest in the world.

Idaho: Craters of the Moon National Monument (20)

This relatively small park covers 50 sq. miles (130 sq. km). It includes, among other features, a lava bed formed only two thousand years ago. The landscape is still barren, and the first life forms are just beginning to colonize the area. There are no tourist facilities in the park, and the only access road is Highway 26/93, between Twin Falls and Arco.

The area can be very interesting for all those visitors who know a little about geology, or have studied recolonization of wild areas by plants and animals after natural disasters.

Among the most interesting birds, the prairie falcon, the

**Idaho:
Snake River Canyon (21)**

sage grouse, many species of sparrows, and the sage thrasher should be mentioned.

This is a large area covering high altitude forests of pine and juniper trees, stretches of sagebrush, and the Snake River Canyon. It covers 1,158 sq. miles (3,000 sq. km). The area is reached from Riggins or Cottonwood by Highway 95 (Idaho), or from Enterprise by Highway 82 (Oregon). In the canyon area there are no roads, and access is possible only on foot or by rafting on the river.

Aside from the magnificent views, this area is interesting because of the number and variety of birds of prey which nest here. At least twelve species of eagles, falcons, hawks, and owls can be seen, often even during a one-day trip.

**Montana:
Bighorn Canyon (22)**

This national recreation area extends over 187 sq. miles (486 sq. km) and includes mountain forests, stretches of sagebrush, and grasslands. It can be reached by a dirt road (Highway 313) from Hardin, in the Crow Indian Reservation, or from Sheridan, Wyoming, by Highway 14. In the canyon which gives this area its name, there are many habitats typical of the eastern slopes of the Rocky Mountains and of the plateaus.

Wapitis, pronghorns, coyotes, red foxes, prairie dogs, and ground squirrels live here. Among the nesting birds, there are sage grouse, magpies, common ravens, lark buntings, longspurs, horned larks, many sparrows, jays, woodpeckers, western meadowlarks, blackbirds, golden eagles, prairie falcons, and so on. Among the reptiles, rattlesnakes, painted turtles, and lizards abound. Rivers are very rich in fish, especially trout.

**Montana:
Missouri Breaks (23)**

This area includes a part of the Missouri River as it flows down from the Rocky Mountains. It is located in north-central Montana. It is possible to visit the mountain forests on many trails and by river rafting. The region is cut through by Highways 236 and 80, as well as by gravel roads. There are also free ferries along the river. Other than this, it is possible to explore the area only on foot or by the river.

The fauna is abundant on the plateaus and in the canyons, especially during the summer. Coyotes, prairie dogs, ground squirrels, pronghorns, and, in some areas, even buffalo can be seen. Among the birds, there are prairie falcons, golden eagles, magpies, sage grouse, longspurs, and

Following page: The Stewart Range is covered by the first winter snow.

Montana: National Bison Range (24)

others. Dangerous rattlesnakes live in the rocky areas.

This famous refuge, covering an area of not even 39 sq. miles (100 sq. km), was created in 1908 in order to protect the buffalo, which at that time numbered a very few animals. It includes stretches of grassland, woody hills, and narrow valleys, and can support about three to five hundred buffalo. Also, many other animals live in the area, like wapiti, mule deer, white-tailed deer, bighorn sheep, pronghorns, badgers, bobcats, and so on. Among the birds are some species which have been introduced from Europe and Asia, like ring-necked pheasants, gray partridges, and chukars, in addition to many local species.

The refuge is reached only by car, following a road about 19 miles (30 km) long, which is open June 1 through September 30.

Montana: Waterton Glacier National Park (25)

This large national park covers over 2,316 sq. miles (6,000 sq. km) of great natural interest. It features massive rock walls, glaciers, snowfields, and high peaks. Access is by the road from Kalispell to West Glacier. Many trails lead up to beautiful alpine meadows, forests, and ridges.

Among the many typical animals and plants living here, the grizzly bear is one. In fact, Glacier is home to one of the largest populations of grizzly bears, numbering perhaps over two hundred. The presence of these animals makes it advisable to take special precautions, especially when camping.

GLOSSARY

alabates pennsylvanica a species of beetle.

alcidae family a group of flying birds including the murre and the tufted puffin.

alluvial plains plains of a river valley.

anadromous migrating from sea water to fresh water to reproduce.

arrete a sharp, steep dividing ridge between cirques, shaped like a knife blade.

brackish a mixture of fresh and salt waters.

carnivorous meat-eating organisms, such as predatory mammals, birds of prey, or insectivorous plants.

castilleja miniata a species of plant also known as "Indian paintbrush."

Cervidae family deer family.

chlorophyll a green pigment contained in leaves required for photosynthesis.

cirque a depression shaped like a large bowl in the earth.

climax fore stone in its final, balanced stage of development.

Compositae family a group of flowering plants.

conifer evergreen tree or shrub with cones that does not drop leaves in the fall.

continental divide mountains in the Rockies from which streams flow east or west.

continental shelf a shallow underwater plain forming a border to a continent, ending with a steep slope.

Corvidae family a group of birds including Clark's nutcracker and the gray jay.

deciduous a plant that sheds its leaves in the fall.

dorsal located on the back.

estivation a period of summer hibernation.

fish ladder a series of higher and higher pools that allows fish to bypass a dam.

fry a fish up to the age of three to five weeks.

Geophilus varians soil centipede.

guano bird droppings.

harem a group of females dominated by a single male.

heliotropism a plant's tendency to follow the path of the sun.

herbivorous plant-eating.

Hylotrechus colonus longhorn beetle.

imprinting a learning process early in the life of a social animal that establishes a behavior pattern.

inseminate to transfer sperm.

invertebrate without a backbone.

jet stream a powerful air current at high altitudes.

kelp a giant seaweed found along coasts.

Lagomorpha order a group of animals including the pica, rabbit, and hare.

metamorphic rock rock changed as a result of pressure and heat into a harder substance.

metamorphosis a fundamental change in structure and/or appearance.

Miocene epoch a period in the earth's history from 15 to 30 million years ago.

moraine a ridge produced by rubble left behind by a glacier.

Mustelidae family a group of marine mammals including the sea otter.

nocturnal active at night.

omnivorous plant- and animal-eating.

parr a fish beyond the age of three to five weeks.

Parulidae family wood warblers.

Pediularis groenlandica a species of plant whose flowers resemble elephant heads.

Penstemon whippleanus a species of plant also known as beardtongue.

photosynthesis the process by which plants combine carbon dioxide with water to make food.

plankton a mixture of tiny or microscopic marine organisms serving as food for various marine species.

Pleistocene epoch a period in the earth's history from 2 million to 5 million years ago.

Plethodon glutinosus slimy salamander.

plumage feathers.

primary forest one that has never been changed by humans.

Prionoxystus robinae carpenter moth.

Quaternary period a period in the earth's history from two million years ago to the present.

rain shadow the area on the leeward side of mountains that receives little precipitation because of its loss on the windward side and at high elevations.

Ranunculus acris a species of buttercup.

Reticulitermes lucifugus termite.

ribbon forest long rows of trees on rounded ridges and plateaus at high elevations.

Romaleum atomarium longhorn beetle.

Sciuridae family a group of mammals including squirrels and chipmunks.

secondary forest one that has regrown.

sedimentary rock rock formed of chemical, organic, or mechanical sediment, such as sandstone and limestone, often transported and deposited by streams.

smolt a salmon or sea trout about two years old.

symmetrical evenly shaped.

travertine a mineral deposit laid down from hot spring waters.

Tremex columba horntail wasp.

Uloma punculatadar kling beetle.

vertebrate with a backbone.

vivaporous producing living young instead of eggs from within the body.

INDEX

Alcidae family, 21
Amphibians, 101
Anemones, 55
Areas of natural interest
  California, 115
  Canada, 105
  Colorado, 115
  Idaho, 118
  Montana, 119
  Oregon, 113
  Washington State, 108
  Wyoming, 115
Arkansas National Wildlife Refuge, 90
Arrete, 19
Avocet, 91, 94

Badger, 83
Banana slug, 44
Banff National Park, 104
Beardtongue, 54
Bears
  black, 47
  grizzly, 65-67, 69, 77, 82
Beaver, 62, 63-65
Beetle, 43
Bering Sea, 24
Bighorn Canyon, 119
Bighorn sheep, 58
Bigleaf maple, 43
Birds
  mountains, 72
  of other wet environments, 85-95
  of prey, 30-31, 95-97
  rain forests, 46
  seabirds, 21
  shore and marshes, 29-31
Black bear, 47
Blackbirds, 90-91
Black-chinned hummingbird, 73
Black-footed albatross, 24
Black-headed grosbeak, 72
Black turnstone, 29
Blue goose, 31
Bobcat, 68, 70
Brackish marshes, 28-29, 30
Broad-tailed hummingbird, 73
Buffalo, 77, 78, 116-117
Buteo, 97
Buttercup, 54, 55

California, areas of natural interest, 115
California sea lion, 24
Calliope hummingbird, 72-73
Canada, areas of natural interest, 105
Canvasback, 90
Cascades, 11-15, 37, 58, 109, 112, 115
Cascades frog, 75
Castillejoa miniata, 54
Cervidae family, 57
Chinook salmon, 34
Chum salmon, 34

Cirque, 19
Clark, William, 7
Clark's nutcracker, 72, 73
Climax forest, 42
Coastal Dunes National Monument, 11, 28
Coho salmon, 34
Colorado, areas of natural interest, 115
Colorado River, 17
Columbia Gorge, 112-113
Columbia National Wildlife Refuge, 112
Columbia River, 14, 15, 16, 17, 25, 35, 112
Common murre, 21-22, 23
Compositae family, 54
Conifers, 37
Continental Divide, 17
Continental shelf, 21
Cook, James, 6
Cormorant, 21, 25
Coyote, 84-85
Crater Lake, 13
Crater Lake National Park, 113-114
Craters of the Moon National Monument, 118-119

Darkling beetle, 43
Deciduous trees, 42
Devil's club, 44
Douglas fir, 40, 42
Dowitcher, 29
Dusky grouse, 46

Elephant's head, 54
Elk, 56
Englemann spruce, 53

Ferruginous hawk, 95
Fir trees, 39
Fish ladders, 35
Fork-tailed petrel, 22
Fraser River, 9, 28
Frogs, 75
Fry fish, 35
Fumaroles, 115

Giant sequoia, 41
Glacier National Park, 9
Glaucous gull, 23
Golden eagle, 97
Golden-manteled ground squirrel, 62
Goshawk, 95
Grand Teton National Park, 115, 118
Grand Teton Range, 16, 88
Grand Tetons, 115
Gray, Robert, 6
Gray jay, 74
Gray's Harbor, 28
Great Basin spadefoot, 101
Greater yellowleg, 29
Great horned owl, 97
Grizzly bear, 65-67, 69, 77, 82

Ground beetle, 43
Ground mollusk, 44
Grouse, 97
Guano, 22
Gulls, 20, 23-24, 25
Gyrfalcon, 31

Haida Indians, 33
Hairy woodpecker, 72
Hawks, 30
Heermann's gull, 23
Heliotropism, 55
Hemlock, 39, 42
Hermit thrush, 72
Heron, 91
Herring gull, 23
Hoary marmot, 61, 62
Hoofed animals, 77
Hummingbirds, 72
Hutton's vireo, 72, 73

Idaho, areas of natural interest, 118-119
Imprinting, 90
Indian paintbrush, 54

Jackrabbit, 81
Jasper National Park, 105
Juan de Fuca Strait, 9

Kelp, 25, 28
Kildeer, 90
Killer whale, 32, 33
Knot, 29

Lagomorpha order, 61
Lamprey, 34, 35
Larches, 53
Lassen Volcanic National Park, 115
Leach's petrel, 22
Leadbetter Point, 28
Least bittern, 91
Lesser yellowleg, 29
Lewis, Meriwether, 7
Liberty Bell Peak, 110-111
Lodgepole pine, 53
Longhorn beetle, 43
Lynx, 70

Madrone tree, 42
Malheur National Wildlife Refuge, 113
Mammals
  hoofed, 77
  of mountains, 55
  predators, 47-51, 65-67, 69, 77, 82-85
  rain forests, 47
  sea mammals, 24
  small animals of plateaus, 79-81
Marbled murrelet, 33
Marmots, 61
Marsh hawk, 95
Miocene period, 37
Missouri Breaks, 119

Montana, areas of natural interest, 119-120
Moose, 57
Moraine, 19
Mosses, 44
Mount Lassen, 115
Mount Ranier, 11
Mount Ranier National Park, 112
Mount Shasta, 11
Mount St. Helens, 12
Mountain beaver, 47
Mountain bluebird, 72, 73-74
Mountain goat, 58-60
Mountain hemlock, 53
Mountain lion, 68, 69-70
Mountains
  birds of, 72
  Cascades, 11-15, 37, 58, 109, 112, 115
  frogs of, 75
  Grand Tetons, 16, 18, 88
  mammals of, 55-72
  Olympic, 9, 40, 53, 58, 69
  predators of, 68
  Sierra Nevada, 115
  Siskiyou, 9
  trees and flowers of, 53
Mule deer, 51-52, 53, 56
Mustelidae family, 26

National Bison Range, 122
North Cascades National Park, 109
Northern alligator lizard, 45
Northern fur seal, 24
Northwestern garter snake, 44-46

Olympic Mountains, 9, 40, 53, 58, 69
Olympic National Park, 108
Oregon, areas of natural interest, 113
Oregon Dunes National Recreation Area, 113
Oregon vine, 29
Osprey, 94

Pacific dogwood, 43
Pacific Rim, British Columbia, 105, 108
Pacific tree frog, 44-45, 75
Parr fish, 35
Parulidae family, 72
Pedicularis groenlandica, 54
Pelicans, 85-87, 89
Penguin, 21
Penstemon whippleanus, 54
Peregrine falcon, 30, 31
Petrels, 22
Pigeon hawk, 30
Pika, 60
Plankton, 21
Plateaus, birds of prey, 95
  grouse, 97
  hoofed animals of, 77
  predators of, 82-85, 95-97

reptiles and amphibians of, 99
small animals of, 79
swans and pelicans of, 85-87, 89
Pleistocene epoch, 9
Plover, 94
Porcupine, 64, 65
Potholes National Wildlife Refuge, 112
Prairie dog, 78, 79-80
Predators
  of mountains, 65
  of plateaus, 82-85, 95-97
Pronghorn, 77-79
Puget Sound, 25, 28, 31-33, 85, 108-109
Purple finch, 72

Quaternary period, 19

Raccoon, 68, 114
Rail, 91, 94
Rain forests
  birds of, 46
  conifers, 37
  deciduous trees, 42
  mammals of, 47
  small animals of, 44
Rain shadow, 15
Ranunculus acris, 54
Rattlesnakes, 99
Red-backed sandpiper, 30
Red-breasted nuthatch, 72, 73
Red fox, 83
Redhead duck, 90
Red-tailed hawk, 31, 94
Red tree vole, 47
Red-winged blackbird, 90
Redwood, 41
Redwood National Park, 115
Reptiles, 99
Rhododendron, 29
Ribbon forest, 53
Ring-necked duck, 90
Rock sandpiper, 30
Rocky Mountain National Park, 115
Rosy finch, 74
Rough-legged hawk, 31, 94
Rough-skinned newt, 45
Ruby-crowned kinglet, 72
Ruddy duck, 90
Ruddy turnstone, 29
Rufous hummingbird, 72

Sage grouse, 97-99
Salal, 29
Salmon, 33-36
Sand dunes, 28-29, 30
Sanderling, 29
Sandhill crane, 88, 89
San Juan Islands, 109
Sciuridae family, 61
Screw mole, 47
Seabirds, 21-24
Seal, 31

127

Sea lion, 20, 24-25, 31
Sea mammals, 24-28
Sea otter, 26-28, 29
Secondary forest, 42
Sharp-tailed grouse, 99
Shearwaters, 24
Shore and marsh birds, 29-31
Shoveler, 92-93
Sierra Nevada Range, 115
Siskiyou Mountains, 9
Sitka spruce, 29, 40
Skagit River, 28, 109
Skagtide Flat Game Refuge, 109
Skunks, 81-82
Slimy salamander, 43
Slugs, 44, 45
Small animals
  plateaus, 79
  rain forests, 44-46
Smolt, 35
Snake River, 16, 17
Snake River Canyon, 119
Snowshoe hare, 71
Soaring hawk, 97
Sockeye salmon, 34
Song sparrow, 46
Sooty shearwater, 24

Sparrow hawk, 94, 95, 97
Spotted owl, 46
Squirrels, 81
Stellar's sea lion, 24-25
Stewart Range, 120
Subalpine fir, 53
Surfbird, 30
Swans, 84, 85-87

Tailed frog, 75
Termite, 43
Tlingit Indians, 33
Townsend ground squirrel, 81
Trees and flowers, mountains, 53
Tufted puffin, 21, 22
Turnstones, 29-30

Vancouver, George, 6
Vancouver Island, 9
Varied thrush, 46

Wandering tattler, 30
Wapiti, 56-57
Washington ground squirrel, 81
Washington State, areas of natural interest, 108-113
Waterton Glacier National Park, 122

Watson Lake, 106-107
Wenatchee, Washington, 112
Western grebe, 90
Western gull, 23
Western hemlock, 40
Western rattlesnake, 99
Western red cedar, 40
Whidbey Island, 109
Whistling swan, 31, 84
White-tailed ptarmigan, 75
Whooping crane, 88, 89, 90
Willamette Valley, 10
Winter wren, 46
Wolf, 70, 71-72, 77, 82
Woodboring beetle, 40
Wood Buffalo National Park, 90
Woodpeckers, 72
Wyoming, areas of natural interest, 115-118

Yellow-bellied marmot, 61, 62
Yellow-headed blackbird, 90
Yellow-shafted flicker, 72
Yellowstone Lake, 87
Yellowstone National Park, 18, 19, 77, 85, 116-118